Biotechnology: DNA to Protein

A Laboratory Project in Molecular Biology

Teresa Thiel
University of Missouri—St. Louis

Shirley Bissen
University of Missouri—St. Louis

Eilene M. Lyons
St. Louis Community College—Florissant Valley

McGraw Hill

Boston   Burr Ridge, IL   Dubuque, IA   Madison, WI   New York   San Francisco   St. Louis
Bangkok   Bogotá   Caracas   Kuala Lumpur   Lisbon   London   Madrid   Mexico City
Milan   Montreal   New Delhi   Santiago   Seoul   Singapore   Sydney   Taipei   Toronto

McGraw-Hill Higher Education

A Division of The **McGraw-Hill** *Companies*

BIOTECHNOLOGY: DNA TO PROTEIN, A LABORATORY PROJECT
IN MOLECULAR BIOLOGY

Published by McGraw-Hill, a business unit of The McGraw-Hill Companies, Inc., 1221
Avenue of the Americas, New York, NY 10020. Copyright © 2002 by The McGraw-Hill
Companies, Inc. All rights reserved. No part of this publication may be reproduced or distributed
in any form or by any means, or stored in a database or retrieval system, without the prior written
consent of The McGraw-Hill Companies, Inc., including, but not limited to,
in any network or other electronic storage or transmission, or broadcast for distance learning.

Some ancillaries, including electronic and print components, may not be available to customers
outside the United States.

This book is printed on recycled, acid-free paper containing 10% postconsumer waste.

International 1 2 3 4 5 6 7 8 9 0 QPD/QPD 0 9 8 7 6 5 4 3 2 1
Domestic   1 2 3 4 5 6 7 8 9 0 QPD/QPD 0 9 8 7 6 5 4 3 2 1

ISBN 0–07–241664–5
ISBN 0–07–112279–6 (ISE)

Publisher: *James M. Smith*
Developmental editor: *Brian S. Loehr*
Associate marketing manager: *Tami Petsche*
Senior project manager: *Marilyn Rothenberger*
Production supervisor: *Enboge Chong*
Coordinator of freelance design: *Michelle D. Whitaker*
Freelance cover/interior designer: *Rokusek Design*
Cover illustration: *Lisa Gravunder*
Supplement producer: *Sandra M. Schnee*
Media technology producer: *Lori A. Welsh*
Compositor: *Lachina Publishing Services*
Typeface: *11/12 Times Roman*
Printer: *Quebecor World Dubuque, IA*

Library of Congress Cataloging-in-Publication Data

Thiel, Teresa.
    Biotechnology: DNA to protein : a laboratory project in molecular biology / Teresa
Thiel, Shirley Bissen, Eilene M. Lyons. — 1st ed.
        p. cm.
    ISBN 0–07–241664–5 — ISBN 0–07–112279–6 (ISE)
    1. Molecular biology —Laboratory manuals. 2. Amylases. 3. Biology projects.
    4. Molecular cloning—Laboratory manuals. I. Bissen, Shirley. II. Lyons, Eilene M.
    III. Title.

QH506 .T48 2002
572.8—dc21   2001030676
                                                            CIP

INTERNATIONAL EDITION ISBN 0–07–112279–6
Copyright © 2002. Exclusive rights by The McGraw-Hill Companies, Inc., for manufacture and
export. This book cannot be re-exported from the country to which it is sold by
McGraw-Hill. The International Edition is not available in North America.

www.mhhe.com

*This manual is dedicated
to the students, who help us
to improve our teaching
by their thoughtful
comments and criticisms.*

About the Authors

Teresa Thiel is a Professor of Biology at the University of Missouri—Saint Louis and serves as Director of the Biotechnology program. She received a bachelor's degree in biology from Virginia Polytechnic Institute and State University and a Ph.D. in microbiology from Case Western Reserve University. Following postdoctoral research at the Plant Research Laboratory at Michigan State University, she joined the faculty at the University of Missouri—Saint Louis. She directs a laboratory of undergraduate and graduate students who work with her on nitrogen fixation and heterocyst development in filamentous cyanobacteria. She has published many papers in this field and has funding for her research from the National Science Foundation and the U.S. Department of Agriculture. She teaches undergraduate and graduate courses in microbiology, virology, and microbial genetics. In addition to these activities, she directs a program for K–12 teachers and students called "Science in the Real World—Microbes in Action." This program, supported by the National Science Foundation, serves to educate teachers and students and promotes an understanding and appreciation of the importance of microorganisms in our world.

Shirley Bissen is an Associate Professor of Biology at the University of Missouri—Saint Louis. She received a bachelor's degree in psychology from the University of Minnesota in Minneapolis and a Ph.D. degree in pharmacology from the University of Michigan in Ann Arbor. She did postdoctoral research at the University of California at Berkeley. Her research focuses on the control of cell division during early embryonic development. She examines these processes in embryos of small freshwater leeches. Each leech embryo undergoes the same pattern of cell divisions, yielding cells that are individually identifiable and whose developmental fates have been characterized extensively. Her research is funded by the National Science Foundation and the March of Dimes Birth Defects Foundation. She teaches undergraduate and graduate courses in molecular biology, biotechnology, and developmental biology. Every summer she helps teach a course on the neurobiology and development of leeches at the Marine Biological Laboratory at Woods Hole, Massachusetts.

Eilene Lyons received her B.S. in biology from Lindenwood University in St. Charles, Missouri, and her M.S. in molecular and cellular biology at the University of Missouri—St. Louis, where she conducted and published research on prokaryotic genetics. She continues to update her biology and biotechnology knowledge and skills at Washington University in St. Louis and by attending NSF-sponsored workshops for community college biotechnology faculty. She taught biology and chemistry at the high-school level, biology at the University of Missouri—St. Louis, and since 1995, biology and biotechnology at St. Louis Community College—Florissant Valley, where she is also the Biotechnology Program Coordinator. She has received numerous awards for excellence in teaching, including Most Dedicated New Faculty, and Outstanding Teacher, presented by students, as well as SLCC Innovator, and Most Outstanding New Teacher of the Year, presented by the faculty.

Contents

Preface

A Project-Based Curriculum

The field of biotechnology has evolved from the principles and applications of molecular biology. Much of biotechnology focuses on the production and characterization of biological molecules, particularly DNA, RNA, and protein. However, like most of biological research, biotechnology typically involves a project that requires the integration of many techniques. Most undergraduate laboratory courses typically involve a series of independent and unrelated laboratory activities that provide information and experience in some particular aspect of the subject. This approach, however, fails to convey how these principles and techniques relate to each other.

This manual presents both the principles of molecular biology that make biotechnology possible and the techniques that come from the practical application of some of these principles. You will characterize the enzyme α-amylase in this project-based course. You will study two important aspects of biotechnology using the α-amylase gene and α-amylase protein: (1) the nature of proteins and their action and (2) the principles and methods for cloning and expressing a gene. Thus, both the analysis of protein and the analysis of DNA are integrated within a single project.

The project begins with a study of the characteristics of the protein α-amylase and its function. You will start by examining the enzymatic action and some commercial uses of α-amylase proteins. Since production of higher yields of α-amylase is the goal of this project, it is logical to understand the protein and the significance of its thermostable characteristic before you begin the work of cloning the gene. Once the commercial importance of this enzyme is established, you will proceed to clone the α-amylase gene of *Bacillus licheniformis* and introduce it into *Escherichia coli,* where the gene will be expressed. At the end of the course, you will return to the study of proteins as you quantitate the production of α-amylase by *E. coli.* In addition, you will use computer modeling software to study the three-dimensional structure of DNA and an α-amylase protein. Exercises in Appendix I will introduce you to some bioinformatics techniques that use sequence databases as well as sequence analysis software.

In this curriculum, you will gain experience in many of the techniques that are important in the field of biotechnology as part of an integrated project. You will begin with a problem, a hypothesis, and a goal that will be achieved by the end of the project. In solving the problem and achieving the goal, you will not only learn the principles of DNA, RNA, and protein structure as well as techniques for analysis of DNA and proteins, but you will also integrate this knowledge as you complete the project. This is particularly apparent near the end of the project when you repeat several earlier performed techniques. This serves to reinforce what you have already learned but also allows you to appreciate how techniques can be used in different contexts to give different information, all of which contribute to the larger overall project.

The final aspect of this curriculum that makes it unique is that the project has a potential practical application in the commercial field of biotechnology. You will be asked to evaluate the success of your project not only in terms of the molecular biology but also in terms of practical considerations such as production and cost.

At the beginning of each lab in this manual, there is a Background section that provides sufficient information for you to understand what you are going to do and why. These sections were not designed to be all-inclusive; your instructor may provide additional information (e.g., alternate detection methods, other thermostable DNA polymerases, etc.). Although not explicitly addressed in the manual, your instructor may have you prepare your own solutions or pour your own media plates (recipes and protocols are provided in the companion Instructor's Manual).

Your Laboratory Notebook

It is important that you keep an organized and accurate record of your activities, results, and observations in a notebook. Your laboratory notebook should contain an up-to-the-minute record of your laboratory activities. This means that you write directly in your notebook, not on separate pieces of paper that you plan to copy into your notebook at a later time. Your notebook is not meant to be a work of art, but it should be legible and organized. Each entry in your laboratory notebook should be dated. Each laboratory exercise should be titled and have the following sections:

Introduction. Briefly explain the purpose of the experiment, i.e., what you are doing and why.

Procedure. This is a record of your laboratory activities, i.e., what materials you used and how you did the experiment. A flow chart diagramming the steps of the experiment is helpful. Provide all of the relevant details, such as volumes, concentrations, time, temperature, the order in which samples were loaded on a gel, etc. Include all of your calculations. (Calculations should be done in your notebook and not in this manual or on a scrap of paper.)

Results. Include all of your experimental results. Present your data in tables, drawings, charts, photographs of gels, photocopies of blots, etc. Throughout this manual, directions regarding your results are provided, e.g., "include this table in your notebook" or "label and place this photograph in your notebook."

Data Analysis. Interpret your results. The Data Analysis sections of this manual tell you what to do. For example, plot your data to generate a standard curve, determine which proteins were detected on your Western blot, estimate the size of the DNA fragments detected via Southern hybridization, determine the number of transformants, etc. Again, show your calculations where relevant.

Summary/Conclusions. Concisely state your major conclusions.

A Sample Lab Schedule

To help you integrate the parts of this project, a sample lab schedule is presented as follows. This sample schedule is for a course that meets twice a week for a semester.

SESSION	LAB #	LAB ACTIVITY
1	Lab 1	Streak bacterial cell plates
2	Lab 1	Streak enzyme, saliva, and detergent plates; analyze all plates
		Micropipetting exercise (Appendix I)
3	Lab 2	Generate maltose standard curve
4	Lab 2	Measure α-amylase activity and quantitate protein
5	Lab 3	Test effects of temperature and pH on α-amylase activity

6	Lab 5A	Prepare samples for SDS-PAGE
7	Lab 4	Protein computer lab
8	Lab 5A	Load and run SDS-polyacrylamide gels; stain one gel with Coomassie Blue
	Lab 5B	Transfer proteins in second gel to membrane
9	Lab 5B	Western blotting
10	Lab 6	DNA computer lab
11	Lab 7	Isolate chromosomal DNA from *Bacillus licheniformis*
12	Lab 7	Check chromosomal DNA on agarose gel
	Lab 8	Assemble PCR reactions to make biotin-labeled α-amylase probe
13	Lab 8	Run PCR products on agarose gel
14	Lab 9A	Assemble restriction cleavage reactions of chromosomal DNA for Southern analysis
15	Lab 9A	Run restriction fragments of chromosomal DNA on agarose gel
	Lab 9B	Denature DNA and set up capillary transfer of DNA to membrane
16	Lab 9B	Affix DNA to membrane
	Lab 9C	Set up Southern hybridization
17	Lab 9C	Detect hybridized DNA
18	Lab 10A	Set up *Hin*d III cleavage of chromosomal DNA
	Lab 10B	Set up *Hin*d III cleavage of plasmid DNA
19	Lab 10A, B	Run cleaved chromosomal DNA and plasmid DNA on agarose gel
	Lab 10C	Assemble ligation reaction
20	Lab 10D	Transformation (spread on agar starch plates)
21	Lab 10E	Primary screening: stain plates, pick colonies, dilute and spread on agar starch plates
22	Lab 10E	Secondary screening: stain plates, pick positive colonies
	Lab 11A	Set up PCR reactions
23	Lab 11A	Run PCR products on agarose gel
	Lab 11B	Set up cultures for mini-prep isolation of plasmid DNA
24	Lab 11B	Isolate plasmid DNA
	Lab 11C	Set up *Bgl* II cleavage reaction (to determine size and orientation of insert)
25	Lab 11C	Run *Bgl* II-cleaved plasmid DNA on agarose gel
	Lab 11C	Set up restriction cleavages of plasmid DNA for mapping
	Lab 11D	Set up cultures for permanent stocks
26	Lab 11C	Run cleaved plasmid DNA on agarose gel
	Lab 11D	Prepare permanent stocks
	Lab 11E	Denature DNA and set up Southern transfer
27	Lab 11E	Affix DNA to membrane and set up Southern hybridization
		Construct plasmid map
28	Lab 11E	Detect hybridized DNA
		Complete plasmid map
29	Lab 12	Inoculate *E. coli* and *Bacillus* cultures
30	Lab 12	Assay enzyme activity of cloned α-amylase gene

Acknowledgments

We would like to thank the many students who have participated in the biotechnology course and have provided valuable comments and criticisms that have helped greatly in producing this book. In particular, Brenda Pratte, an undergraduate research student at the time these materials were developed, deserves special thanks for her efforts in developing many aspects of the project. We thank Jeff Elhai, who provided the positive selection vector, pRL498, that was critical to the success of cloning the α-amylase gene.

The development of this project-based biotechnology course would not have been possible without the support of the National Science Foundation. We deeply appreciate the funding they provided for this project through the Course and Curriculum Development Program and the Instrumentation and Laboratory Improvement Program (DUE-9354731 and DUE-9452142, respectively). In addition, we appreciate the materials, equipment, and supplies generously provided by Sigma-Aldrich Chemical Company to support this program.

List of reviewers

Steven McCommas
Southern Illinois University—Edwardsville

Jeffrey D. Newman
Lycoming College

Stephen Randall
Indiana University/Purdue University—Indianapolis

Christina Sax
University of Maryland—University College

Danielle R. Tilley
Central Seattle Community College

Marcelo Tolmasky
California State University—Fullerton

Cornelius A. Watson
Roosevelt University

INTRODUCTION TO THE PROJECT

Why Do We Need a Good Source of α-Amylase?

BACKGROUND

Starch is a storage polymer of glucose that is abundant in plants. Two polymers make up starch: amylose, which is a linear chain of glucose units joined by α-1,4 glycosidic bonds, and amylopectin, a branched polymer that is similar to amylose except that its branches are attached to the backbone via α-1,6 linkages. Agricultural crops that are not used for food are used to produce large amounts of starch for the manufacture of adhesives, food additives, and sugar syrups.

Sugar syrups that are made from starch include glucose, maltose, fructose, and sorbitol. Since fructose is more than twice as sweet as sucrose (cane sugar or beet sugar is mostly sucrose), many processed foods and beverages are sweetened with fructose. High concentration fructose syrup, made from cornstarch, is the most economically important of the syrups made from starch.

Although the glycosidic bonds of starch are readily broken by acidic conditions (pH 1.4–2.0) at high temperatures (140–150°C), such conditions produce undesirable colored by-products. The removal of these by-products requires expensive purification steps to produce syrups suitable for use in foods and beverages. Over the last 20–30 years, the production of syrups by enzymatic treatment of starch has replaced much of the acid treatment. Starch treated with enzymes yields sugars with fewer by-products and thus requires little further purification.

Since no single enzyme can convert starch to fructose, the production of fructose syrup is a multistep process that requires several enzymes. These steps are outlined in Figure I.1. First, in the gelatinization step, the starch is mixed into an aqueous solution and heated to dissolve the starch. Second, in the liquefaction step, α-amylase from *Bacillus licheniformis* is added to break the α-1,4 glycosidic linkages of starch, producing shorter chains of glucose (oligosaccharides). Then, additional enzymes are used to digest the oligosaccharides into smaller sugar units. Two of the major commercial products of starch conversion are maltose syrup and fructose syrup, each produced by the action of microbial enzymes on oligosaccharides made from starch. Note, however, that the different enzymes used in the processing of starch have different temperature and pH requirements. This makes the processing of starch to syrup a complex process.

One of the advantages of using the α-amylase from *B. licheniformis* for the liquefaction step is that this enzyme is very heat-stable (thermostable) and can be used immediately after gelatinization, which is done at high temperature. This enables the gelatinization and liquefaction steps to be done very quickly, without any changes or additions to the starch slurry.

FIGURE I.1

Production of maltose and fructose syrups from starch. In step 1, the gelatinization step, the starch is mixed into an aqueous solution and heated to dissolve the starch to a gel-like consistency. In step 2, the liquefaction step, *α*-amylase is added to the solubilized starch to digest the long chains of glucose (starch) into shorter chains of glucose (oligosaccharides). In step 3, the saccharification step, the oligosaccharides are further digested with either (*a*) amyloglucosidase, to yield a glucose syrup or (*b*) a different *α*-amylase, to yield a syrup high in maltose. In step 4, the isomerization step, glucose is partially converted to fructose by the enzyme glucose isomerase, yielding a high fructose syrup.

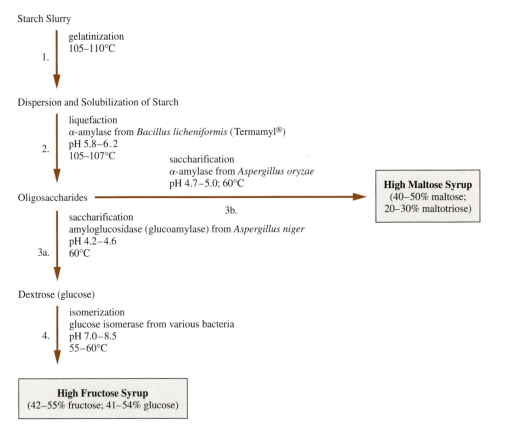

THE PROBLEM

One problem with the enzymatic method of starch degradation is the need to produce large amounts of bacterial enzymes. The *α*-amylase from *B. licheniformis* is an excellent enzyme for the liquefaction step because it is thermostable. The problem, however, is that *B. licheniformis* does not produce large amounts of the enzyme.

THE GOAL

The goal of this project is to increase the production of the *α*-amylase that is normally made by *Bacillus licheniformis*. Since it is economical to add the *α*-amylase to the hot starch solution, an *α*-amylase that is active at high temperatures is very desirable. The approach that you will take is to genetically engineer the bacterium *Escherichia coli* to produce large amounts of this enzyme. *E. coli* is widely used in the biotechnology industry for the large-scale production of many proteins that were originally made by other organisms. For example, human insulin is produced by *E. coli* as a result of genetic engineering techniques. In order to accomplish this goal, the gene that encodes the *α*-amylase of *B. licheniformis* must be transferred to *E. coli*, where it must be expressed well enough to produce large amounts of the *α*-amylase protein. This recombinant *E. coli* strain will then be a potentially economical source of a thermostable *α*-amylase for the processing of starch to syrups.

EXERCISE

Formulate a Hypothesis

Most scientific research comes from efforts to answer a question or solve a problem. Scientists often state their question of interest in the form of a hypothesis that can be tested. Based on the information provided and the goal you have been given, formulate a hypothesis for this project on which you will be working for the next few months. At the end of this project, you will review this hypothesis and determine whether or not your hypothesis was correct.

Biotechnology industries are interested in profit. At the end of the project, you will also have to determine whether you think the approach you have used is a cost-effective way to make more thermostable α-amylase and whether the project would be considered successful from the perspective of profit-making potential.

LAB 1

Starch Plate Assay

GOALS

The goals of this lab are to examine the degradation of starch by the enzyme α-amylase and to explore some sources and commercial uses of this enzyme.

OBJECTIVES

After completing Lab 1, you will be able to explain or describe

1. the biochemical reaction catalyzed by α-amylase
2. several sources of α-amylase
3. secretion of α-amylase by some bacteria
4. a commercial use of α-amylase
5. the role of α-amylase in that product

BACKGROUND

Enzymes are proteins that function as organic catalysts in cells. They speed up metabolic reactions over a millionfold, allowing the biochemical reactions that are essential for life. Many enzymes are produced commercially for use in the biotechnology industry. Alpha-amylase is an enzyme that is commercially important for the production of corn syrup from cornstarch, as an additive in laundry detergents, for sizing (coating) in the textile industry, in the brewing of beer, and in the production of bread. Because of the large amounts of α-amylase needed for these industrial applications, this is an economically important enzyme.

Starch and glycogen are storage polymers of glucose made via dehydration synthesis reactions between the glucose molecules. Starch is found in plants and contains a mixture of two glucose polymers, amylose and amylopectin. Amylose is a linear chain of glucose molecules linked between carbon 1 of one molecule and carbon 4 of another molecule (α-1,4 glycosidic linkage) (Fig. 1.1). Amylopectin is similar to amylose except it also has branches that are linked between carbon 1 of a glucose molecule in one chain and carbon 6 of a glucose molecule in another chain (α-1,6 glycosidic linkage). The ratio of amylose to amylopectin in starch is about 1:3 or 1:4. Glycogen is found in animal cells; its structure resembles that of amylopectin except glycogen is more highly branched.

The enzyme α-amylase catalyzes the hydrolysis of α-1,4 glycosidic linkages in starch (or glycogen) to yield maltose, maltotriose, and α-dextrin (Fig. 1.2). Maltose has two glucose residues in an α-1,4 linkage, and maltotriose has three glucose residues linked α-1,4. Alpha-dextrin consists of an α-1,4 backbone and

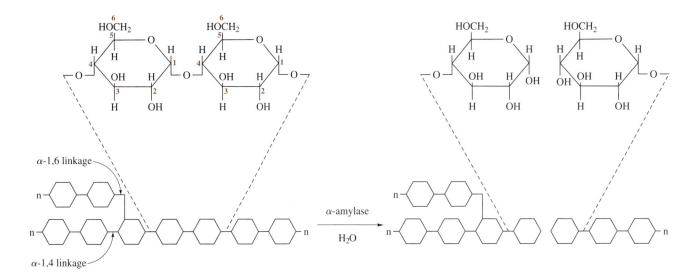

FIGURE 1.1

Alpha-amylase hydrolyzes α-1,4 glycosidic linkages. An amylopectin component of starch is shown on the left, with a region enlarged to show the α-1,4 glycosidic linkage between two glucose residues. Upon hydrolysis, shown on the right, carbons 1 and 4 have hydroxyl groups, and the glucose polymer is broken into shorter chains. The letter n indicates that there are more sugar residues that are not shown in the figure.

associated α-1,6 branch points. The enzyme β-amylase, found in barley seeds, also hydrolyzes starch by cleaving successive maltose units from the ends of the chains.

All cells synthesize enzymes that are used for metabolic processes within the cell. Enzymes catalyze the synthesis and hydrolysis of complex organic molecules such as starch, proteins, fats, and nucleic acids. Some cells, but not all, are able to secrete enzymes. Secretion is a mechanism that some cells use to transport enzymes through the cell membrane and into the surrounding environment. For example, enzymes may be secreted to hydrolyze organic molecules outside the cell. The products of this hydrolysis can then be transported across the cell membrane and into the cell for further metabolism.

In mammals, α-amylase is made in the salivary glands and is secreted into saliva. Although stomach acids destroy salivary α-amylase, the pancreas makes more α-amylase and secretes it into the intestines, where it participates with other

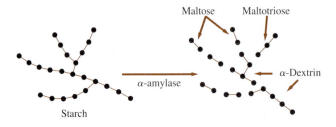

FIGURE 1.2

Products of starch hydrolysis by α-amylase. Alpha-amylase hydrolyzes internal α-1,4 glycosidic bonds to produce maltose, maltotriose, and α-dextrin. Maltose contains two glucose residues with an α-1,4 linkage, whereas maltotriose contains three glucose residues with α-1,4 bonds. Alpha-dextrin contains several glucose residues with both α-1,4 and α-1,6 glycosidic bonds.

digestive enzymes in the breakdown of starch and glycogen to glucose. Many bacteria and fungi also produce α-amylase, which they typically secrete into the environment to break down starch so that smaller sugars can enter the microbial cell. Alpha-amylase is stored in the seeds of plants and is activated upon germination to break down the starch in the seed that leads to the production of glucose for the plant embryo.

LABORATORY OVERVIEW

In this lab, you will examine the enzymatic action of α-amylase. You will streak commercially purified α-amylase, saliva, various detergents (Part I), and different strains of bacteria (Part II) onto LB agar plates that contain starch. (Although LB agar plates contain a nutrient-rich medium for the growth of bacterial cells, these plates also provide a convenient support for the other solutions.) The assay for α-amylase activity will be the degradation of starch, which will be monitored with an iodine color test. Starch reacts with the iodine in Lugol Solution to produce a blue-black color. If the starch in the agar plate is hydrolyzed by α-amylase, there will be no reaction with iodine, and this area of agar will appear "clear."

TIMELINE

Part I can be done in one lab session. It takes about 15–30 minutes to streak the plates, 30 minutes to incubate the plates, and about 30 minutes to view the plates. Part II requires two lab sessions. It takes about 15 minutes to streak the plates, which must be incubated overnight, and then it takes about 30 minutes to analyze the plates.

SAFETY GUIDELINES

You will be working with live bacterial cultures in this lab. Read the section on Sterile Technique presented in Appendix II before starting this lab.

Discard the tubes of bacteria and any saliva-contaminated waste into bags that will be autoclaved prior to disposal.

Part I: Assay Samples of Purified α-Amylase, Saliva, and Detergents

PROCEDURE

STREAK THE PLATES

□ 1. Obtain two LB agar + starch plates, and divide each into two parts by drawing a line across the bottom of the plate with a permanent marking pen. Label one half of each plate as "enzyme" and the other half as "saliva." Add one student's name or initials to one plate and the other student's name or initials to the second plate. Each student will streak enzyme and saliva on her or his plate.

□ 2. Read the section on Proper Enzyme Usage in Appendix II; keep the tube of enzyme on ice. Soak a cotton swab in the tube of α-amylase enzyme purified from *Bacillus licheniformis* and streak over the "enzyme" half of your plate (Fig. 1.3).

FIGURE 1.3

Diagram for streaking enzyme, saliva, and detergent streak plates. (*a*) Enzyme/saliva plate. (*b*) Detergent plate.

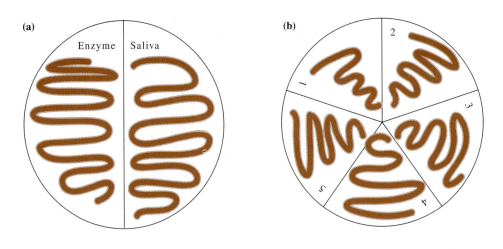

3. Place the cotton tip of a cotton swab under your tongue and allow it to become soaked with saliva. Streak this swab over the "saliva" half of your plate (Fig. 1.3). Discard the swab in an autoclave bag.

4. Obtain another LB agar + starch plate, and divide it into five sections by drawing lines on the bottom. Write the brand name of one of the detergents you are testing on each section, and label the plate with your group number or initials.

5. Streak each section of this plate (see Fig. 1.3) with a cotton swab soaked in the corresponding 10% detergent solution. Start in the center of the plate and streak outward.

6. Incubate all four plates at 37°C for about 30 minutes.

ANALYZE THE PLATES

1. After the 30-minute incubation period, gently flood the surface of each enzyme/saliva plate with about 2–3 ml of Lugol Solution. Gently rotate each plate until the entire surface of the plate is covered with Lugol Solution. Wait for 1–2 minutes. The iodine in the Lugol Solution will react with the starch in the agar, changing the color of the agar-starch medium from yellow to dark blue-black. After about 2 minutes, pour off and discard the Lugol Solution and replace the lids. Hold the plates up to the light or against a white piece of paper, and observe them from the bottom rather than from the top.
Note: The staining of starch with iodine fades rather quickly. You may want to outline the areas of clearing on the bottom of the plate with a marking pen so that you can examine the zones of clearing even after the stain disappears.

2. Examine each enzyme/saliva plate, looking for zones of clearing. In your notebook, draw and label the results seen on these plates. Shade the areas where there is an iodine-starch reaction.
Alternatively, you can photograph the plates. First, you may have to restain them with Lugol Solution. Then, place a plate on a white light transilluminator and use a Polaroid camera (with a yellow filter) to take a photograph. The camera settings will depend upon the type of film used.

3. Examine the detergent plate. The areas on which the detergents were streaked will turn pinkish in color, in contrast to the dark blue-black of the starch zones. This pinkish color is due to interactions between surfactants and other additives in the detergents and the agar medium

FIGURE 1.4
Iodine-stained starch plate streaked with detergents.

(it is not due to α-amylase activity). Some of the detergent areas will "clear" and become lighter or paler in color (but may not clear to the color of the LB agar). Pay attention to the order in which the detergent areas clear. Some clear quickly, some clear after several minutes, and some never clear (Fig. 1.4). Record your results in your notebook. Make a drawing of this plate, labeling each section and shading areas where there was an iodine-starch reaction. Alternatively, take a photograph of the plate.

DATA ANALYSIS

- Analyze the results observed on these plates by completing Table 1.1. Estimate the extent of clearing on each plate (column 3), and rank the solutions according to the amount of α-amylase activity (column 4). Include this table in your notebook.
- Analyze the results observed on the detergent plate by completing Table 1.2. Explain the criteria you used to rank the detergents (column 4). Include this table in your notebook.

TABLE 1.1			
Activity of Purified α-Amylase and Saliva			
Plate	**Solution Tested**	**Degree of Clearing (No clearing = 0) (Extensive clearing = ++++)**	**Rank of Solutions (Most activity = 1)**
1	Purified α-amylase (from *B. licheniformis*)		
	Saliva 1		
2	Purified α-amylase (from *B. licheniformis*)		
	Saliva 2		

Detergent Tested	Degree of Clearing (No clearing = 0) (Extensive clearing = +++)	α-Amylase Present? (Yes or no)	Rank of Detergents (Most activity = 1)

TABLE 1.2
Enzyme Activity of Detergents

Part II: Assay Bacterial Cells

PROCEDURE

STREAK THE PLATES

☐ 1. Read the section on Sterile Technique in Appendix II, if not done already. Obtain two more LB agar + starch plates, and label the bottom of one as "*B. amyloliquefaciens*" and the other as "*B. licheniformis*." Add your group number or initials to each plate.

☐ 2. Gently vortex your tube of *Bacillus amyloliquefaciens* liquid culture to resuspend the cells. If you are using a metal inoculating loop, sterilize it by putting it in a flame until it is red. Allow it to cool for about 20 seconds, and then touch the loop to the agar along the edge of the plate; the agar may sizzle if the loop is still too hot. Using sterile technique, dip the sterile loop into the tube of liquid culture and streak the culture of *B. amyloliquefaciens* over one-fourth of the appropriate plate (Fig. 1.5). Sterilize the loop by flaming, and streak through the first quadrant onto the second quadrant. (Pass your loop through the first streaks about four or five times and then continue to streak without going through the original streaks.) Flame the loop, and streak from the second quadrant to the third quadrant, etc. (Fig. 1.5). This method is called *quadrant streaking* and is designed to isolate individual colonies on the plate.

☐ 3. Gently vortex your tube of *B. licheniformis* liquid culture. Sterilize the inoculating loop again by flaming it until it is red. Using sterile technique, streak the other plate with the *B. licheniformis* liquid culture, using the quadrant streak method to produce individual colonies.

☐ 4. Incubate these plates (inverted) at 37°C overnight. *Note:* You or your instructor will remove these plates from the incubator and refrigerate them until the next laboratory session.

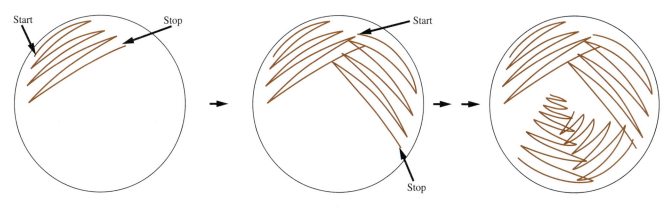

FIGURE 1.5
Diagram for quadrant streaking.

ANALYZE THE PLATES

☐ 1. Before staining these plates, observe the pattern and density of bacterial growth on them. The center of the colonies should be thick and dense, whereas the edges should be much thinner. You may want to outline the edges of the colonies on the bottom of the plate with a marking pen so that later you will not confuse thin cell growth with clearing.

☐ 2. Flood each plate with about 2–3 ml of Lugol Solution. Gently rotate the plate until the entire surface is covered with Lugol Solution. Wait about 2 minutes, and then pour off and discard the Lugol Solution. Hold the plates up to the light or against a white background and observe them from the bottom rather than from the top.

Since the staining of starch with Lugol Solution fades rather quickly, you may want to outline the areas of clearing on the bottom of the plate with a marking pen (of a different color) so that you can examine the zones of clearing after the stain disappears.

☐ 3. Examine the plates, looking for zones of clearing. *Note:* The cell colonies prevent the Lugol Solution from reacting with starch directly under the area of bacterial growth. Do not confuse bacterial growth with clearing. Look for clearing beyond the edges of the colonies.

☐ 4. In your notebook, draw and label these plates, indicating the areas of bacterial growth and the areas of clearing. Shade the areas where there was an iodine-starch reaction. Alternatively, you can photograph the plates; label and place the photo in your lab notebook.

TABLE 1.3			
Enzyme Activity of *Bacillus* Cells			
Cells Tested	**Degree of Clearing (No clearing = 0) (Extensive clearing = ++++)**	**α-Amylase Secreted?**	**α-Amylase Produced?**
B. amyloliquefaciens			
B. licheniformis			

DATA ANALYSIS

- Analyze your results by completing Table 1.3. Place this table in your lab notebook.

QUESTIONS

1. Why do some plates streaked with saliva show greater or lesser degrees of clearing?
2. Is either species of the bacterium *Bacillus* a good source of α-amylase? Explain your answer.
3. Explain your *Bacillus licheniformis* data. How can it be that your plate of *B. licheniformis* cells (in Part II) showed little or no clearing, but the commercial preparation of α-amylase purified from *B. licheniformis* (in Part I) showed extensive clearing?
4. Among the detergents that appeared to contain α-amylase, was there a difference in the extent of clearing? Give one reason why some detergents might hydrolyze starch better than other detergents.
5. Check the labels on the detergent containers.
   a. Are enzymes listed as an ingredient in any of the detergents used? If so, which detergents?
   b. Is α-amylase listed as an ingredient on any of the detergents?
   c. Is there a correlation between detergents that show clearing and whether they contain enzymes? If so, give an explanation.
6. You are a biochemist working for a company that produces laundry detergents, and you are interested in obtaining large amounts of α-amylase for your product.
   a. What are two possible sources of α-amylase for your product?
   b. Which source would you use, and why do you think that would be the best source?
   c. In the Sigma Chemical Company catalog, α-amylase from human saliva costs 1000 times more than the same amount of α-amylase from the bacterium *Bacillus*. Why might it cost so much more? Does this change your answer to the previous question? Explain.

LAB 2

Quantitative Enzyme Assay

GOALS

The goals of this lab are to measure the enzyme activity and determine the specific activity of α-amylase in saliva and in a solution of commercially purified α-amylase.

OBJECTIVES

After completing Lab 2, you will be able to

1. construct a standard curve for known amounts of maltose
2. determine the enzyme activity of solutions that contain α-amylase
3. construct a standard curve for known amounts of protein
4. determine the specific activity of salivary α-amylase and purified *Bacillus licheniformis* α-amylase

BACKGROUND

ENZYME FUNCTION

Enzymes are organic catalysts that are usually proteins. They speed up metabolic reactions over a millionfold. Although energetically favorable, metabolic reactions do not occur under normal physiological conditions unless enzymes are present. In order for molecules to react, they must collide with a minimum amount of energy, called the *activation energy* (E_a). This required energy can be thought of as an energy hill that molecules must climb. In metabolic reactions, this hill is so high that reactions rarely happen (Fig. 2.1). Enzymes lower the activation energy, allowing reactions to occur many times more frequently. Enzymes are not consumed in reactions but rather are used over and over. They work by chemically interacting with substrate molecules at a cleft or pocket in the enzyme called the *active site*. Enzymes are very specific, acting only on those substrates that fit precisely into their active sites. The molecules of the active site interact chemically with those of the substrate sitting in the active site. These interactions strain bonds within the substrate, making it easier to break old bonds and form new bonds. Once the reaction has occurred, the product is released, and the enzyme can bind another substrate molecule. A unit of enzyme activity generally is defined as the amount of enzyme that will consume 1 μmole of substrate or liberate 1 μmole of product per minute under optimum conditions.

FIGURE 2.1

Enzymes accelerate reactions by lowering the activation energy.

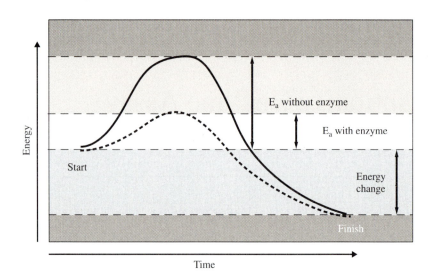

QUANTITATIVE α-AMYLASE ASSAY

In Lab 1, you assayed α-amylase activity by the disappearance of its substrate (starch). In this lab, you will measure α-amylase activity by the appearance of a product (maltose) of the reaction (Fig. 1.2). As α-amylase catalyzes the hydrolysis of the α-1,4 linkages in a solution of starch, the amount of maltose increases. Maltose can be detected after reacting with 3,5-dinitrosalicylic acid, a component of the Maltose Color Reagent (Bernfeld 1951, Bernfeld 1955). This compound absorbs more light at a particular wavelength after reacting with maltose. To quantitate the amount of maltose produced by salivary or commercial preparations of α-amylase, the absorbance of known amounts of maltose that have reacted with Maltose Color Reagent will first be determined and graphed to produce a standard curve.

This α-amylase assay consists of two separate steps. First, the enzyme is allowed to react with saturating amounts of starch under specific conditions (i.e., pH 7 at room temperature). Second, Maltose Color Reagent is added to the sample so that the amount of maltose generated during the enzyme reaction step can be calculated from the standard curve. From these data, you will be able to calculate the units of α-amylase activity per ml of sample. For the purposes of this project, one unit of α-amylase is defined as the amount of enzyme needed to produce 1 mg of maltose from starch in 1 minute at room temperature and at pH 7.

ENZYME SPECIFIC ACTIVITY

The specific activity of an enzyme is defined as the number of units of enzyme activity per mg of protein. If an enzyme is pure, the amount of enzyme activity per mg of protein (i.e., the specific activity) is high; if the solution contains other proteins, the specific activity is lower. The mass of protein in a sample includes all of the proteins, not just the enzyme of interest.

To determine the specific activity of α-amylase, the total amount of protein will be measured using the Protein Determination Reagent, which contains bicinchoninic acid, a compound that absorbs more light at a particular wavelength after reacting with protein. First, a protein standard curve will be generated by measuring the absorbance of known amounts of albumin protein after reaction with the Protein Determination Reagent. Second, the amount of total protein in samples of saliva and purified *B. licheniformis* α-amylase will be determined. Lastly, the specific activity of each sample will be determined by dividing its units of α-amylase activity by its total mass of protein.

LABORATORY OVERVIEW

In Part I of this lab, a standard curve for maltose will be generated. In Part II, this maltose standard curve will be used to determine the amount of α-amylase activity in samples of saliva and a commercial preparation of *B. licheniformis* α-amylase. In Part III, a standard curve for protein will be constructed so that the total mass of protein in the samples can be determined and the specific activity of the different preparations of α-amylase can be calculated.

TIMELINE

This lab can be carried out over two lab sessions. Part I takes about 1 hour, Part II takes about 1.5–2 hours, and Part III takes about 1 hour.

SAFETY GUIDELINES

Handle your own saliva samples only. Dispose of any saliva-contaminated pipettes, tips, and tubes in autoclave bags.

Wear gloves when handling the Maltose Color Reagent (1% 3,5-dinitrosalicylic acid, 0.4 M NaOH, 1.06 M sodium potassium tartrate) as this concentration of NaOH is slightly corrosive. Flush your skin or eyes with water if there is contact.

Part I: Generate a Standard Curve for Maltose

PROCEDURE

☐ 1. Read about using automatic micropipettes in the Background section of the Micropipetting Exercise in Appendix I, and test your pipetting skills by performing the exercise before starting this lab.

☐ 2. Label six round-bottom, 15-ml tubes from "1" to "6." Add the appropriate volumes of water and maltose, listed in Table 2.1, to each tube. Add 1.0 ml of Maltose Color Reagent to each tube. Cap and vortex to mix. ***WEAR GLOVES WHEN HANDLING THE MALTOSE COLOR REAGENT.***

☐ 3. Place the tubes in a dry heat block set at 100°C (or a boiling water bath) for exactly 15 minutes. At this temperature, the 3,5-dinitrosalicylic acid in the Maltose Color Reagent reacts with maltose to produce compounds that absorb more light at 540 nm.

TABLE 2.1 Assay Volumes						
Tube	1 (Blank)	2	3	4	5	6
Deionized water	2.0 ml	1.8 ml	1.6 ml	1.4 ml	1.2 ml	1.0 ml
Maltose (2 mg/ml)	0	0.2 ml	0.4 ml	0.6 ml	0.8 ml	1.0 ml
Maltose Color Reagent	1.0 ml	1.0 ml	1.0 ml	1.0 ml	1.0 ml	1.0 ml

Tube	1 (Blank)	2	3	4	5	6
A_{540nm}	0					
mg of maltose	0					

TABLE 2.2
Maltose Standards

☐ 4. Place the tubes on ice until cooled to room temperature. Add 9 ml deionized water (dH_2O) to each tube. Tightly cap the tubes, and mix the contents by inverting the tubes several times. (Since the tubes are very full, vortexing does not mix the contents very well.)

☐ 5. Using tube 1 (contains no maltose) as the blank to zero the spectrophotometer, measure the absorbance at 540 nm (A_{540nm}) for tubes 2–6. Record the results in Table 2.2. Include this table in your notebook.

DATA ANALYSIS

- Calculate the mass (mg) of maltose in each sample and record the value in Table 2.2. Mass (mg) of maltose can be calculated by using the information given in Table 2.1. For example,

  0.2 ml × 2 mg/ml = 0.4 mg maltose

- Plot the data in Table 2.2 on graph paper, putting the total mass (mg) of maltose on the x-axis and the A_{540nm} on the y-axis. Use a ruler to draw a straight best-fit line through the origin (0,0) and the data points. This line is the standard curve that will be used to determine the amount of maltose present in a solution on the basis of its absorbance at 540 nm. This standard curve will also be used in Labs 3 and 12. Include this graph in your notebook.

Part II: Measure α-Amylase Activity

PROCEDURE

DILUTE ENZYME SAMPLES

You will assay α-amylase activity in samples of saliva and a commercial preparation of α-amylase purified from *Bacillus licheniformis*. These samples must be diluted a thousandfold prior to the α-amylase assay. Enzyme solutions should always be kept on ice prior to use, as the cold temperature slows down any inactivation processes (see the section on Proper Enzyme Usage in Appendix II).

☐ 1. Prepare two 50-ml tubes for diluting the saliva samples. Label one tube as "saliva 1" and the other as "saliva 2." Add 40 ml of dH_2O to each tube, and store on ice.

2. One student will contribute the saliva 1 sample and the other student will contribute the saliva 2 sample. ***HANDLE ONLY YOUR OWN SALIVA SAMPLE.*** Collect some saliva by placing a plastic disposable bulb-type pipette under your tongue and gently squeezing the bulb to draw saliva into the pipette. Transfer the saliva to a 1.5-ml microcentrifuge tube.

3. Use scissors to cut the tip off a small pipette tip so that the opening is wider. Use this tip to remove 40 μl of saliva from the 1.5-ml microcentrifuge tube. Saliva is very viscous and difficult to pipette accurately. To facilitate this process, hold the saliva tube at eye level and watch to make sure that saliva is drawn into the tip while you slowly release the plunger. Keep the tip immersed in the saliva for several seconds after releasing the plunger to make sure that the entire volume is drawn into the tip. Add the 40 μl of saliva to the appropriate dilution tube. Again, visually monitor that the saliva is released from the pipette tip. You may have to flush out the tip with the dilution water by pipetting up and down several times. Cap each tube and shake vigorously to mix. Store on ice.

4. Label another 50-ml tube for diluting the purified α-amylase. Add 40 ml of dH$_2$O, and store the tube on ice. Your instructor will give each group a tube of concentrated α-amylase that was purified from *Bacillus licheniformis.* (Keep this tube of enzyme on ice.) Add 40 μl of the concentrated α-amylase to the dilution tube. The concentrated α-amylase solution contains 50% glycerol, which is viscous and difficult to pipette accurately. Therefore, pipette carefully and watch to make sure that the correct volume of enzyme solution is transferred from one tube to the next. Shake to mix, and store on ice.

ENZYME REACTION STEP

1. Label seven round-bottom, 15-ml tubes as "blank," "saliva 1A," "saliva 1B," "saliva 2A," "saliva 2B," "α-amylase A," and "α-amylase B." The "A" and "B" tubes are duplicates that should be treated identically, using exactly the same sample material. The blank is a control tube; it will receive everything except the enzyme.

2. Add 1 ml of dH$_2$O to the blank tube.

3. Add 1 ml of diluted saliva 1 to each of the appropriate duplicate assay tubes. Add 1 ml of diluted saliva 2 to each of the appropriate duplicate assay tubes. Add 1 ml of the diluted purified α-amylase to each of the appropriate duplicate assay tubes.

4. Add 1 ml of 1% starch, pH 7, to each tube, including the blank tube. Cap and vortex to mix. Let each tube sit at room temperature for exactly 12 minutes. Start timing when starch is added to the first tube. During this reaction period, α-amylase hydrolyzes α-1,4 linkages of starch to yield maltose and other products.

MALTOSE DETECTION STEP

1. ***WEAR GLOVES WHEN HANDLING THE MALTOSE COLOR REAGENT.*** In the same order that the starch was added, add 1 ml of the Maltose Color Reagent to each tube. Cap, vortex, and place in a dry heat block set at 100°C (or a boiling water bath) for exactly 15 minutes. Start timing when the Maltose Color Reagent is added to the first tube. During this step, the Maltose Color Reagent reacts with the maltose that was produced during the enzyme reaction step.

TABLE 2.3
Enzyme Activity of Salivary and Purified *B. licheniformis* α-Amylases

Test Solution	Saliva 1A	Saliva 1B	Saliva 2A	Saliva 2B	α-Amylase A	α-Amylase B
A_{540nm}						
mg of maltose produced during assay[1]						
Calculated mg of maltose produced by 1 ml of starting test solution[2]						
Units of enzyme in 1 ml of starting test solution						
Average units of enzyme for duplicate samples						

[1]Take into account any dilutions made when measuring the A_{540nm} of the samples.
[2]Take into account the original thousandfold dilutions of the saliva and α-amylase.

2. Place the tubes on ice until cooled to room temperature. Add 9 ml of dH_2O to each tube. Tightly cap the tubes and invert several times to mix.

3. Using the blank tube to zero the spectrophotometer, measure the absorbance (A_{540nm}) for the six sample tubes and record the values in Table 2.3. If the A_{540nm} of your saliva sample is higher than the values on your maltose standard curve (from Part I), then you must dilute your sample (and the blank) with dH_2O and read the absorbance again. If you do this, record how much you diluted the samples in your notebook.

DATA ANALYSIS

- Calculate the mass (mg) of maltose that was produced in each of your samples during the enzyme reaction step. Use your maltose standard curve generated in Part I for this. From the A_{540nm} value on the y-axis, find the corresponding mg of maltose on the x-axis. Record these values in Table 2.3.
- If you diluted your saliva samples when measuring the A_{540nm}, then you must multiply the amount of maltose read from your standard curve by your dilution factor to determine how much maltose was produced during the enzyme reaction step. For example, suppose you diluted your sample three-fold (one part of sample and two parts of dH_2O) and got an A_{540nm} reading of 0.4. From your standard curve, you determined that this reading corresponded to 1.2 mg of maltose. Thus, the total amount of maltose produced during the enzyme incubation step would be

1.2 mg × 3 (dilution factor) = 3.6 mg of maltose

- Calculate and record in Table 2.3 the mg of maltose that would have been produced by 1 ml of each of the starting test solutions during the enzyme incubation step. Since you added 1 ml of a thousandfold diluted solution to the assay, you measured enzyme activity in 1/1000 of a milliliter or 1 μl of the starting test solution. Thus,

mg of maltose/1 μl × 1000 μl/ml

= mg of maltose/1 ml of starting solution

- Calculate and record in Table 2.3 the units of enzyme activity in 1 ml of each of the starting test solutions. A unit of α-amylase enzyme activity is the amount of enzyme needed to liberate 1 mg of maltose from starch in 1 minute at room temperature and at pH 7. Since the enzyme incubation step lasted 12 minutes, you must divide the calculated mg of maltose (Table 2.3, row 4) by 12 to determine the amount of maltose produced in 1 minute.
- Determine and record in Table 2.3 the mean number of units of enzyme for each of your duplicate samples. Include this table in your notebook.

Part III: Quantitate Protein

PROCEDURE

ASSEMBLE SAMPLES

☐　1. In the second column of Table 2.4, the final amount of the protein standard (albumin) is listed for each tube. Calculate and record the volume (in μl) of a 1 mg/ml solution of albumin to add to each tube to give the desired amount of albumin. (Volume = concentration $\times$ mass.) Calculate and record the volume (in μl) of water to add to each tube to bring the final volume to 100 μl. Sample calculations are presented as follows:

1 mg/ml $\times$ 0.02 mg = 0.02 ml $\times$ 1000 μl/ml = 20 μl

100 μl $-$ 20 μl of albumin = 80 μl of dH$_2$O

☐　2. Label six test tubes from "1" to "6." Using the values you entered in Table 2.4, add the appropriate volumes of water and albumin stock solution to each tube.

☐　3. Table 2.5 lists the components to be added to your experimental and control samples for the protein assay. You will measure the amount of protein in your undiluted starting saliva and the concentrated α-amylase. You need a water control tube for the saliva samples. You also need a glycerol control tube for the α-amylase sample because the α-amylase is prepared in a solution of 50% glycerol, and glycerol itself interacts with the Protein Determination Reagent.

TABLE 2.4
Albumin Standards

Tube Number	Mass of Albumin in Assay	Volume of 1 mg/ml Albumin Stock	Volume of dH$_2$O	A$_{562}$
1 (blank)	0	0	100 μl	0
2	0.020 mg	20 μl	80 μl	
3	0.040 mg			
4	0.060 mg			
5	0.080 mg			
6	0.100 mg			

	TABLE 2.5 Experimental Samples			
Condition	Volume of Test Solution	Volume of dH₂O	Volume of 50% Glycerol	A_{562}
H₂O blank	0	100 μl	0	0
Undiluted saliva 1	25 μl	75 μl	0	
Undiluted saliva 2	25 μl	75 μl	0	
Glycerol blank	0	0	100 μl	0
Concentrated α-amylase	100 μl	0	0	

4. Label five more test tubes, as listed in Table 2.5. Retrieve the tubes of undiluted saliva collected in Part II of this lab (or collect fresh samples if Part III is not done on the same day as Part II) and your tube of concentrated α-amylase. Using pipette tips with cutoff ends, add the appropriate volumes of undiluted saliva and concentrated enzyme to each tube. Add the appropriate volumes of dH₂O or 50% glycerol to each tube.

PROTEIN REACTION STEP

1. Add 2 ml of Protein Determination Reagent to each tube (the six tubes listed in Table 2.4 and the five tubes listed in Table 2.5). Cap and vortex to mix.
2. Incubate all the tubes at 37°C for 30 minutes.
3. Using the tube without albumin (tube 1) as the blank, measure and record in Table 2.4 the absorbance at 562 nm (A_{562nm}) for the albumin standards (tubes 2–6).
4. Using the water control tube as the blank, measure and record in Table 2.5 the A_{562nm} for both saliva samples. Using the glycerol control tube as the blank, measure and record in Table 2.5 the A_{562nm} for the α-amylase sample.

DATA ANALYSIS

CALCULATE PROTEIN MASS

- Using the data in Table 2.4, plot the mass (mg) of albumin on the x-axis and A_{562nm} on the y-axis. Draw a straight, best-fit line through the origin and the data points. This is your albumin standard curve. Include this graph in your notebook.
- Use your albumin standard curve to determine the amount (mg) of protein in each of the experimental samples, and record these in Table 2.6.
- Calculate and record in Table 2.6 the total amount of protein in 1 ml of each starting test solution. Take into account the volume of each test solution added to the protein assay. For example, you determined the amount of protein in 25 μl of saliva. Convert this to 1 ml as follows:

$$\text{mg of protein}/25 \ \mu\text{l} \times 1000 \ \mu\text{l/ml} = \text{mg protein}/1 \ \text{ml of saliva}$$

- Perform a similar calculation for the concentrated α-amylase sample. Include Table 2.6 in your notebook.

CALCULATE SPECIFIC ACTIVITY

- Calculate and record in Table 2.7 the specific activity of α-amylase (units of enzyme per mg of protein) in each test solution. The specific activity of α-amylase is the total units of enzyme activity in 1.0 ml of sample (from Table 2.3) divided by the total protein in 1.0 ml of sample (from Table 2.6). Share these data with other class members by recording Table 2.8 and Table 2.9 on the board for all to copy. Include all these tables in your notebook.

TABLE 2.6 Total Protein in Saliva and Commercially Purified α-Amylase			
Test Solution	Saliva 1	Saliva 2	Purified α-Amylase
mg of protein present in sample tested (from standard curve)			
mg of protein present in 1 ml of original test solution			

TABLE 2.7 Specific Activity of Salivary and Purified *B. licheniformis* α-Amylases			
Test Solution	Saliva 1	Saliva 2	Purified α-Amylase
Units of enzyme per ml (from Table 2.3)			
Total mg of protein per ml (from Table 2.6)			
Specific activity (units of enzyme per mg of protein)			

TABLE 2.8 Class Data of Specific Activity of Salivary α-Amylase										
Student #	1	2	3	4	5	6	7	8	9	10
Units of α-amylase per ml										
Total mg of protein per ml										
Specific activity (units/mg)										
Student #	11	12	13	14	15	16	17	18	19	20
Units of α-amylase per ml										
Total mg of protein per ml										
Specific activity (units/mg)										

Group #	1	2	3	4	5	6	7	8	9	10
Units of α-amylase per ml										
Total mg of protein per ml										
Specific activity (units/mg)										

TABLE 2.9
Class Data of Specific Activity of Purified *B. licheniformis* α-Amylase

QUESTIONS

1. Suppose you do an enzyme assay of a solution suspected of having α-amylase activity. The solution is mixed with a starch solution, incubated, and tested for maltose using the Maltose Color Reagent. Why are you testing for maltose?

2. If there were maltose already in the solution that you were testing for α-amylase activity (question 1) before you added any starch, would this cause a problem? Why? If so, how could you correct for this problem in the assay?

3. Why are duplicate samples (A and B) prepared for each sample in Part II?

4. Compare and contrast the results from Lab 1 and Lab 2 in terms of sensitivity of the assay, ease of performing the assay, and information obtained in each assay.

5. You have been hired by a chemical company to identify 1000 individuals that produce large amounts of α-amylase in their saliva from a population of 10,000 test subjects. Your budget is $40,000. If the starch plate test costs $2.00 per plate and the quantitative α-amylase assay using the Maltose Color Reagent costs $10.00 per test, how would you screen the 10,000 people? Describe the most economical method for the selection of the 1000, while still determining quantitative α-amylase activity (and spending all of the $40,000).

6. In Lab 1, starch agar plates streaked with saliva showed greater or lesser areas of clearing. Based on the data in Table 2.8, do some people have more α-amylase in their saliva? Explain.

7. Examine the class data in Table 2.8 and Table 2.9. How does the specific activity of α-amylase in saliva compare with the specific activity of the commercially purified α-amylase? What can you conclude about these two preparations of α-amylase?

LAB 3

Factors Affecting Enzyme Function

GOAL

The goal of this laboratory is to investigate the effects of temperature and pH on the activity of α-amylase enzymes from three different sources.

OBJECTIVES

After completing Lab 3, you will be able to

1. assay α-amylase activity at different temperatures and pHs
2. graph enzyme activity data
3. analyze enzyme activity data to determine the optimal temperature and pH for a given enzyme
4. use that information to identify the source of the enzyme

BACKGROUND

Proteins are polymers of amino acids, linked by covalent peptide bonds. The chain of amino acids folds into a complex three-dimensional shape that is maintained by weak, noncovalent interactions between nonadjacent amino acid residues. These interactions include hydrogen bonds that form when a hydrogen atom is shared by two other atoms of the peptide bonds and/or amino acid side chains, electrostatic interactions between positively and negatively charged side chains, and hydrophobic interactions between nonpolar amino acid side chains. Disulfide bonds between cysteine residues in the chain can also stabilize protein structure. The final three-dimensional shape of a polypeptide depends upon the sequence of amino acids in the chain. It is the three-dimensional structure that allows proteins to interact with substrate molecules and catalyze reactions. Enzymes work by chemically interacting with substrate molecules at a cleft or pocket in the enzyme called the *active site.* The active site is formed by the three-dimensional conformation of the protein and is complementary in shape and charge to the substrate. If the shape or ionic properties of the active site are modified, the substrate may not bind correctly and the enzyme may not function properly, thus slowing or preventing the reaction.

Every enzyme has an optimum temperature and pH range at which its structure is most stable, the substrate binds best, and the enzyme is most active. For example, digestive enzymes secreted in the stomach function best at acidic pH, while those secreted in the small intestine work well at neutral pH. Some enzymes, such as those found in hot springs bacteria, are active at high temperatures, while those found in bacteria from the Arctic work at low temperatures. At temperature and pH values other than the optimum, enzymes have less activity or no activity. Changes in temperature or pH disrupt the weak interactions that stabilize the three-dimensional structure of the protein. For example, changes in pH can alter the ionization of side chains and disrupt the electrostatic interactions involved in proper protein folding. Likewise, perturbation of the ionization of side groups of the amino acids in the active site can affect substrate binding and catalytic activity. More extreme changes in pH or increased temperatures can lead to complete unfolding or denaturation of the protein and complete loss of enzyme function.

LABORATORY OVERVIEW

In this lab, you will be given one of three samples of α-amylase, each from a different source. One source is the bacterium *Bacillus licheniformis,* whose α-amylase functions well at high temperatures and acidic pH. Although the α-amylase of *Bacillus licheniformis* functions best at these extremes, it retains activity over a broad range of temperature and pH values. Another source is the fungus *Aspergillus oryzae,* whose α-amylase functions optimally under warm temperatures and moderately acidic conditions. The third source is porcine (pig) pancreas, whose α-amylase has a narrow zone of activity; it functions best at body temperature (37°C) under neutral conditions. After determining the optimal temperature (Part I) and the optimal pH (Part II) for your unknown sample, you will identify the source of the α-amylase in your sample.

Your instructor will provide you with a tube of concentrated α-amylase that is labeled A, B, or C. Keep enzymes on ice to minimize their inactivation; refer to the section on Proper Enzyme Usage in Appendix II. The α-amylase is provided as a concentrated stock solution because it is more stable when it is stored at a high concentration in the presence of glycerol. You will dilute the enzyme a hundred-fold prior to the enzyme assay of each part of the lab.

TIMELINE

This lab is divided into two parts; each takes about 1 hour.

SAFETY GUIDELINES

Wear gloves when handling the Maltose Color Reagent (1% 3,5-dinitrosalicylic acid, 0.4 M NaOH, 1.06 M sodium potassium tartrate) as this concentration of NaOH is slightly corrosive. Flush your skin or eyes with water if there is contact.

Part I: Effect of Temperature on α-Amylase Activity

PROCEDURE

ENZYME REACTION STEP

☐   1.   Record in your notebook which α-amylase preparation (A, B, or C) you are using. In this part of the lab, you will measure α-amylase activity at a range of temperatures from 4°C to 100°C. Label one

□   round-bottom, 15-ml tube as "blank" and five other 15-ml tubes as "4°C," "23°C," "37°C," "65°C," and "100°C." Add 1 ml of 1% starch solution, pH 7, to each tube.

□   2.   Place each tube in a water bath or incubator that is set to one of the indicated temperatures. The "blank" and "23°C" tubes should be left at room temperature; the "4°C" tube can be put in a bucket of ice. Allow the tubes to equilibrate to the desired temperature for 10 minutes.

□   3.   Make a fresh dilution of your α-amylase sample (A, B, or C) by mixing 0.1 ml (100 μl) of the concentrated enzyme stock solution with 9.9 ml of dH_2O. To ensure that you measure the concentrated enzyme solution accurately, pipette this glycerol-containing mixture slowly and carefully. Vortex to mix. Keep your stock and diluted enzyme mixtures on ice.

□   4.   Retrieve all the assay tubes after the 10-minute preincubation step. Add 1 ml of dH_2O to the blank tube only. Cap and vortex to mix. Add 1 ml of diluted α-amylase to the other five tubes. Cap and vortex to mix. Return each tube to its appropriate temperature for exactly 12 minutes. Start timing when α-amylase is added to the first tube. During this time, α-amylase hydrolyzes α-1,4 linkages in starch to produce maltose and other products.

MALTOSE DETECTION STEP

□   1.   ***WEAR GLOVES WHEN HANDLING THE MALTOSE COLOR REAGENT.*** In the same order that the starch solution was added, add 1 ml of Maltose Color Reagent to each of the tubes, including the blank. Cap, vortex, and place in a dry heat block set at 100°C (or a boiling water bath) for exactly 15 minutes. Start timing when the Maltose Color Reagent is added to the first tube.
Note: The 0.4 M NaOH in the Maltose Color Reagent denatures all of the enzymes.

□   2.   Place the tubes on ice until cooled to room temperature. Add 9 ml of dH_2O to each tube. Tightly cap the tubes and invert several times to mix.

□   3.   Using the blank to zero the spectrophotometer, measure the A_{540nm} of all the samples. Record these results in Table 3.1. If an absorbance reading is higher than the values on your maltose standard curve (from Lab 2), then you must dilute that sample and the blank with dH_2O and read them again. Make note of this dilution in your notebook.

DATA ANALYSIS

- Use the maltose standard curve from Lab 2 to determine the amount of maltose produced at each temperature. Record these values in Table 3.1. If any of the samples were diluted during the absorbance reading step, this

TABLE 3.1 α-Amylase Activity as a Function of Temperature					
Temperature	4°C	23°C	37°C	65°C	100°C
A_{540nm}					
mg of maltose produced					

TABLE 3.2 Class Data of α-Amylase Activity at Different Temperatures					
Temperature	4°C mg of maltose	23°C mg of maltose	37°C mg of maltose	65°C mg of maltose	100°C mg of maltose
α-amylase A					
α-amylase B					
α-amylase C					

must be taken into account when calculating the mg of maltose produced during the enzyme reaction step. Include Table 3.1 in your notebook.

- Graph the results, placing temperature on the x-axis and mg of maltose produced on the y-axis. Include this graph in your notebook.
  Note: You cannot determine units of enzyme activity because you performed the α-amylase assay at temperatures other than room temperature, as specified in the standard definition of units of α-amylase activity.
- Share your data with other class members by recording Table 3.2 on the board for all to copy. Determine the optimal temperature for each of the three samples of α-amylase.
- Can you identify the source of your sample of α-amylase?

Part II: Effect of pH on α-Amylase Activity

PROCEDURE

ENZYME REACTION STEP

1. In this part of the lab, you will use the same preparation of α-amylase (A, B, or C) and measure enzyme activity at several different pH levels. Label one round-bottom, 15-ml tube as "blank" and five other 15-ml tubes as "pH 5," "pH 6," "pH 7," "pH 8," and "pH 9." Add 1 ml of 1% starch of the appropriate pH to the corresponding assay tube. You must use a different starch solution for each tube. Add 1% starch, pH 7, to the blank tube.

2. Make a fresh dilution of your α-amylase preparation by mixing 0.1 ml (100 μl) of the concentrated enzyme stock with 9.9 ml of dH$_2$O. The concentrated enzyme solution contains glycerol; pipette it slowly and carefully. Vortex to mix. Keep your stock and diluted enzyme mixtures on ice.

3. Add 1 ml of dH$_2$O to the blank tube only. Cap and vortex. Add 1 ml of dilute α-amylase solution to each of the other tubes. Cap and vortex to mix. Incubate all six tubes at room temperature for exactly 12 minutes. Start timing when α-amylase is added to the first tube. During this incubation step, α-amylase hydrolyzes α-1,4 linkages in starch, yielding maltose and other products.

MALTOSE DETECTION STEP

1. ***WEAR GLOVES WHEN HANDLING THE MALTOSE COLOR REAGENT.*** In the same order that starch solution was added, add

TABLE 3.3 α-Amylase Activity as a Function of pH					
pH	pH 5	pH 6	pH 7	pH 8	pH 9
A_{540nm}					
mg of maltose produced					

TABLE 3.4 Class Data of α-Amylase Activity at Different pHs					
pH	pH 5 mg of maltose	pH 6 mg of maltose	pH 7 mg of maltose	pH 8 mg of maltose	pH 9 mg of maltose
α-amylase A					
α-amylase B					
α-amylase C					

1 ml of the Maltose Color Reagent to each of the tubes, including the blank. Cap, vortex, and place in a dry heat block set at 100°C (or a boiling water bath) for exactly 15 minutes. Start timing when the Maltose Color Reagent is added to the first tube.

2. Place the tubes on ice until cooled to room temperature. Add 9 ml of dH_2O to each tube. Tightly cap the tubes and invert several times to mix.

3. Using the blank to zero the spectrophotometer, measure the A_{540nm} of all the sample tubes. Record the absorbance of each sample in Table 3.3. If an absorbance reading is higher than the values on your maltose standard curve (from Lab 2), you must dilute that sample and the blank with dH_2O and read them again. Make note of this dilution in your notebook.

DATA ANALYSIS

- Use the standard curve from Lab 2 to determine the amount of maltose produced under each condition. Record these values in Table 3.3. If any of the samples were diluted during the absorbance reading, then you must take into account the dilution factor when calculating the mg of maltose produced during the enzyme reaction step. Include Table 3.3 in your notebook.
- Graph the results, placing pH on the x-axis and mg of maltose produced on the y-axis. Include this graph in your notebook.
  Note: You cannot determine units of enzyme activity because you performed the α-amylase assay at pH values other than pH 7, as specified in the standard definition of units of α-amylase activity.
- Share your data with other class members by recording Table 3.4 on the board. Determine the optimal pH for each of the three samples of α-amylase.
- Identify the source of your α-amylase. Examination of the data of other groups will help you determine the source of your α-amylase.

QUESTIONS

1. Based on the characteristics of the enzymes presented in the Laboratory Overview section of this lab and the data from this lab (see Table 3.2 and Table 3.4), identify the source of each of the α-amylases (A, B, and C).

2. What temperature produced the lowest activity for each of the enzymes? Explain, in terms of enzyme function and the effects of temperature on that function, why perishable foods are stored at 4°C.

3. In the production of high fructose corn syrup from cornstarch, the first step is to solubilize or dissolve the starch by mixing it with an aqueous solution and heating the mixture to about 105°C. The next step is the digestion of the starch into shorter oligosaccharides with the enzyme α-amylase. Addition of α-amylase directly to the hot starch solution would minimize the time (and hence, the cost) of the process. Which of the three enzymes tested in this lab would be best for this application? Why?

4. A commercial laundry washes heavily soiled chefs' aprons in 65°C water with a strong alkaline detergent (the pH of the diluted detergent and water mixture is about 8.5). You have been asked to determine whether it would be feasible to add α-amylase to this detergent to improve the cleaning of the aprons. Based on the results of your experiments here, do you think it is feasible, and if so, which enzyme source or sources would you suggest? Justify your answer.

5. Design a Maltose Color Reagent assay using the detergent provided by the commercial laundry (see question 4) to determine whether your chosen α-amylase would function well in the presence of the alkaline detergent. Indicate what should be added to the experimental tube and the control tubes (there should be three control samples). Indicate the conditions for the enzyme incubation step of the assay.

LAB 4

Analysis of Protein Structure Using RasMol

GOAL

The goal of this lab is to examine all levels of protein structure and, in particular, the structure of an α-amylase protein using the computer program RasMol.

OBJECTIVES

After completing Lab 4, you will be able to

1. identify which amino acid residue contributes the carboxyl group and which contributes the amino group to a peptide bond
2. determine the approximate mass of a given polypeptide
3. identify and discuss the four levels of protein structure
4. use the RasMol computer software to visualize a given protein and identify the components of its three-dimensional structure
5. use the RasMol computer software to visualize an α-amylase protein and identify its active site, inhibitor, and associated ions

BACKGROUND

PROTEIN STRUCTURE

Proteins are polymers, or chains, whose subunits comprise a particular class of organic molecules called *amino acids.* There are typically 20 different amino acids that form amino acid polymers. These amino acids are joined together by strong covalent bonds called *peptide bonds,* and the protein polymers are called *polypeptides.* A peptide bond forms between the carboxyl group on one amino acid and the amino group on the other (Fig. 4.1). The symbol "R" represents the side chains of the amino acids. This polypeptide chain constitutes the *primary structure* of the protein, and the particular amino acid composition of each polypeptide determines the final biochemical characteristics of the protein. Polypeptide chains can have only a few amino acid residues or as many as several thousand amino acid residues. The mean molecular weight of an amino acid is 110; thus, the molecular weight of a polypeptide that has 1000 amino acid residues would be about 110,000. This is usually expressed in units called *daltons,* which are equal to one atomic mass unit. Therefore, the mass of a protein with 1000 amino acid residues is about 110,000 daltons, or 110 kd (kilodaltons).

FIGURE 4.1
Formation of a peptide bond.

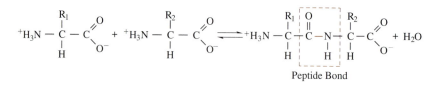

Peptide Bond

Polypeptides do not exist as long, straight chains, however. Rather, each polypeptide chain folds into a distinct three-dimensional shape or conformation, which is determined by its primary amino acid sequence. Chemical interactions between amino acids, which may be near each other but may also be distant from each other in the primary sequence, contribute to the folding of a protein. The regular repeating patterns of twists or kinks of a polypeptide chain constitute the protein's *secondary structure.* Two common secondary structures are the α-helix and the β-sheet. These structures arise from regular hydrogen bonding between carbonyl and amino groups of particular amino acid residues within a contiguous stretch of polypeptide chain.

Additional interactions between nonadjacent amino acid side chains cause the protein to fold into other three-dimensional shapes. This third level of protein structure is called the *tertiary structure.* In addition to hydrogen bonds, there are electrostatic interactions between charged side chains and hydrophobic interactions between nonpolar side chains. For example, amino acid residues with hydrophobic side chains will generally interact with each other to avoid the aqueous environment in which they are dissolved. Conversely, amino acids with hydrophilic side chains will tend to remain on the outside of the protein, hydrogen bonding with the water molecules of the solution. All these weak, noncovalent bonds contribute to the specific shape a protein assumes. Covalent bonds can also stabilize the protein structure. The amino acid cysteine contains sulfur, and the sulfurs in two cysteine residues can form a covalent bond called a *disulfide bridge.*

In addition to chemical interactions within a polymer, there may also be chemical interactions between two or more polypeptide chains. These interactions between different polypeptides allow proteins to form more complex structures, called the *quaternary structure.* Like the bonds that form secondary and tertiary structures, the bonds that form the quaternary structure are typically weak chemical bonds. The chemical bonds that form the quaternary structure are between amino acids on different polypeptide chains.

The α-amylase protein you will examine is from pig (porcine) pancreas; it consists of a single polypeptide chain containing 496 amino acid residues. The amino acid sequence of porcine α-amylase is provided in Fig. 4.2. Each amino acid residue is identified by its position in the polypeptide chain (e.g., gln1, tyr2, ala3, pro4, etc.). A polypeptide chain is always written starting with the terminal amino acid that has a free amino group, called the *amino terminal end of the protein,* and ends with the terminal amino acid that has a free carboxyl group, called the *carboxyl terminal end of the protein.*

ENZYME FUNCTION

Enzymes are usually proteins that function as metabolic catalysts; they enhance the rate of a reaction by lowering the activation energy. Reactions between molecules occur only when substrate atoms collide with some minimum amount of energy. Enzymes act to decrease this necessary energy by interacting with substrate molecules at the active site on the enzyme. The binding of the substrate to the active site is very specific and precise. Particular amino acids in the active site interact chemically with the substrate, straining the bonds between substrate atoms and making it easier to break and make bonds.

gln	tyr	ala	pro	gln	thr	gln	ser	gly	arg	thr	ser	ile	val	his
leu	phe	glu	trp	arg	trp	val	asp	ile	ala	leu	glu	cys	glu	arg
tyr	leu	gly	pro	lys	gly	phe	gly	gly	val	gln	val	ser	pro	pro
asn	glu	asn	val	val	val	thr	asn	pro	ser	arg	pro	trp	trp	glu
arg	tyr	gln	pro	val	ser	tyr	lys	leu	cys	thr	arg	ser	gly	asn
glu	asn	glu	phe	arg	asp	met	val	thr	arg	cys	asn	asn	val	gly
val	arg	ile	tyr	val	asp	ala	val	ile	asn	his	met	cys	gly	ser
gly	ala	ala	ala	gly	thr	gly	thr	thr	cys	gly	ser	tyr	cys	asn
pro	gly	ser	arg	glu	phe	pro	ala	val	pro	tyr	ser	ala	trp	asp
phe	asn	asp	gly	lys	cys	lys	thr	ala	ser	gly	gly	ile	glu	ser
tyr	asn	asp	pro	tyr	gln	val	arg	asp	cys	gln	leu	val	gly	leu
leu	asp	leu	ala	leu	glu	lys	asp	tyr	val	arg	ser	met	ile	ala
asp	tyr	leu	asn	lys	leu	ile	asp	ile	gly	val	ala	gly	phe	arg
ile	asp	ala	ser	lys	his	met	trp	pro	gly	asp	ile	lys	ala	val
leu	asp	lys	leu	his	asn	leu	asn	thr	asn	trp	phe	pro	ala	gly
ser	arg	pro	phe	ile	phe	gln	glu	val	ile	asp	leu	gly	gly	glu
ala	ile	lys	ser	ser	glu	tyr	phe	gly	asn	gly	arg	val	thr	glu
phe	lys	tyr	gly	ala	lys	leu	gly	thr	val	val	arg	lys	trp	ser
gly	glu	lys	met	ser	tyr	leu	lys	asn	trp	gly	glu	gly	trp	gly
phe	met	pro	ser	asp	arg	ala	leu	val	phe	val	asp	asn	his	asp
asn	gln	arg	gly	his	gly	ala	gly	gly	ser	ser	ile	leu	thr	phe
trp	asp	ala	arg	leu	tyr	lys	val	ala	val	gly	phe	met	leu	ala
his	pro	tyr	gly	phe	thr	arg	val	met	ser	ser	tyr	arg	trp	ala
arg	asn	phe	val	asn	gly	glu	asp	val	asn	asp	trp	ile	gly	pro
pro	asn	asn	asn	gly	val	ile	lys	glu	val	thr	ile	asn	ala	asp
thr	thr	cys	gly	asn	asp	trp	val	cys	glu	his	arg	trp	arg	glu
ile	arg	asn	met	val	trp	phe	arg	asn	val	val	asp	gly	glu	pro
phe	ala	asn	trp	trp	asp	asn	gly	ser	asn	gln	val	ala	phe	gly
arg	gly	asn	arg	gly	phe	ile	val	phe	asn	asn	asp	asp	trp	gln
leu	ser	ser	thr	leu	gln	thr	gly	leu	pro	gly	gly	thr	tyr	cys
asp	val	ile	ser	gly	asp	lys	val	gly	asn	ser	cys	thr	gly	ile
lys	val	tyr	val	ser	ser	asp	gly	thr	ala	gln	phe	ser	ile	ser
asn	ser	ala	glu	asp	pro	phe	ile	ala	ile	his	ala	glu	ser	lys	leu

Alpha-amylase enzymes are part of a superfamily of enzymes that have a distinctive conformation called an $(\alpha/\beta)_8$ barrel. In these proteins, eight parallel strands of β-sheet coil to form a central barrel, which is surrounded by eight α-helices (Fig. 4.3). The active site of α-amylase lies at one end of the barrel, and five glucose residues of a starch polymer fit into the active site (cleavage occurs between the third and fourth residues). Alpha-amylase also has a chloride ion and a calcium ion that are important for its function. The calcium ion helps the protein fold and maintain its shape while the chloride ion interacts with the substrate at the active site.

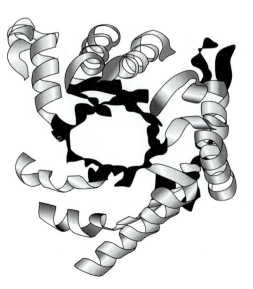

FIGURE 4.3
The $(\alpha/\beta)_8$ barrel region of α-amylase. Beta-strands are shown in black, and α-helices are shown in gray.

BACKGROUND QUESTIONS

1. Give the abbreviated name and position number of the amino acid residue at the amino terminal end and the amino acid residue at the carboxyl terminal end of porcine α-amylase (see Fig. 4.2).
2. Amino acids pro4 and gln5 of α-amylase form a peptide bond. Which amino acid contributes the amino group and which contributes the carboxyl group of the peptide bond between them?
3. The amino acid cysteine contains sulfur. The sulfurs in two cysteine residues can form a covalent bond called a disulfide bridge that contributes to the tertiary structure of a protein. Identify, by number, two amino acids in porcine α-amylase that might form a disulfide bridge.
4. What is the approximate mass of porcine α-amylase?
5. Define *active site*. Describe the active site of α-amylases.

LABORATORY OVERVIEW

RasMol is a program that converts the crystallographic data used to determine the three-dimensional structure of a protein into a computer image. This public domain software is available at <http://www.umass.edu/microbio/rasmol/>. The protein files required for this lab can be downloaded from <http://www.rcsb.org/pdb> or from this manual's website at <http://www.mhhe.com/thiel>. The script file, which contains specific information regarding the α-amylase protein, must be downloaded from this manual's website (see preceding address).

The RasMol program uses a combination of menu items (chosen with the mouse) and typed commands to view particular aspects of protein structure. In Parts I, II, and III, you will explore the primary, secondary, and tertiary structure of crambin, a very small protein from plants. In Parts IV, V, and VI, you will analyze the structure, active site, and disulfide bridges in the porcine α-amylase protein.

Two other protein analysis programs, Chime and Protein Explorer, are also available at <http://www.umass.edu/microbio/rasmol/>. Both are RasMol derivatives and display macromolecular structures in three dimensions.

TIMELINE

This computer lab takes 2–3 hours; it can be done in class or assigned as homework.

PROCEDURE

USING A MACINTOSH TO VIEW PROTEINS WITH RASMOL

1. The RasMac icon has three circles arranged in a triangle. Follow directions given by your instructor for opening the program from the desktop.
2. There should be two windows: one with a black background titled "RasMol," the other a white window titled "RasMol Command Line." Adjust the size of the windows by clicking on and dragging (up, down, left, right) the small box on the lower right-hand corner of each window. The white RasMol Command Line window should cover about one-third of the left of the screen, and the black RasMol window should cover about two-thirds of the right side of the screen.
3. Go to step 4 of the PC instructions that follow.

USING A PC TO VIEW PROTEINS WITH RASMOL

1. From the Start Menu, open RasMol. The black "RasMol" window with the menu items above it will be displayed.
2. On the Task Bar at the bottom of the screen, the item "RasMol Command" will be displayed. Open this by clicking on it. This will open a white window with a prompt line that accepts typed commands.
3. Adjust the size of the two windows so that both can be seen at the same time. You may need to move the RasMol Command Line window slightly to the left in order to reduce it from the right side. Move the mouse to the outside line of the window that you want to change until you see a double-ended arrow. Click and hold the mouse while you drag the side of the window to the size desired. The RasMol Command Line window should cover about one-third of the left of the screen, and the black RasMol window should cover about two-thirds of the right side of the screen.
4. In the instructions that follow, the left side of the instruction table presents the commands that should be typed in the RasMol Command Line window. The right side of the table comments on (explains) the purpose of each command. Type only the command, not the comment. After typing the command, push the "Enter" key on the keyboard. *TO MINIMIZE ERRORS, TYPE COMMANDS CAREFULLY, AND MAKE SURE SPACES ARE PLACED ONLY WHERE SHOWN IN THE COMMANDS. BULLETED ITEMS DIRECT YOU TO USE THE MOUSE TO CHOOSE MENU ITEMS OR MANIPULATE THE IMAGE.*
5. By the end of these exercises, you will be very familiar with many of the RasMol commands. If you want more information, consult the RasMol User Manual, available from the Help menu. You can also type "help," followed by the topic name in the Command Line window, to get information presented there.

Part I: Primary Protein Structure

Commands—type text below and press "enter"	Comments
load crambin.pdb	loads the *protein* *data* *bank* (pdb) file for crambin

- View the crambin protein using the different options under the Display menu on the black "RasMol" window. A good choice from the Color menu is "structure."

1. Which view option in the Display menu gives the most chemical detail?
   _____

2. Which view option gives the best display of the secondary structure, such as helices (corkscrew-like structure)? _____
   _____

3. Which view option gives the most realistic display in terms of the actual relative size of the atoms of the molecule? _____ _____

- Select "ball & stick" from the Display menu and "cpk" from the Colors menu.

You now have a chemical view of crambin. Every knob represents an atom, and the sticks represent covalent bonds. Each atom is colored by chemical convention. In cpk, gray = carbon; red = oxygen; blue = nitrogen; and yellow = sulfur (hydrogens are not shown). You can use the mouse to click on an atom. The Command Line will tell you the atom and the molecule in which the atom is located. The atom letters are C, O, N, and S (sometimes these letters are followed by other letters, such as A, B, D1, etc.). The molecule is also called the "group" and has the three-letter abbreviation for the amino acid and its number. The amino acids in crambin will now be studied in more detail.

restrict leu	displays only leucine residues

- Use the mouse to click on each atom and identify it.
  4. How many carbons are there in leucine? How many oxygens? How many nitrogens? How many sulfurs? _____ _____

- Next you will view the arginine residue next to leucine 18.

select arg17	selects arginine 17

- To view the molecule, choose "ball & stick" from the Display menu. Use the mouse to click on each atom and identify it.
  5. How many carbons are there in arginine? How many oxygens? How many nitrogens? How many sulfurs? _____ _____

- Next you will look at both arg17 and leu18; note that they are not joined by a peptide bond. The following commands will join them.

select arg17, leu18	selects arginine 17 and leucine 18

- To see the peptide bond form, you need to have a good view of both molecules and a good view of the small gap between them. Rotate the molecules until there is a good view of the gap.
- To see the peptide bond form, choose "ball & stick" from the Display menu and watch the gap between the two amino acids disappear as they are joined by a peptide bond. (If you did not see it, go back to the command "restrict leu" and try again.)
  6. Which two specific atoms (and from which amino acids) were joined to make the peptide bond? _____ _____

select all	selects all the amino acids

- Choose "ball & stick" from the Display menu.

Families of amino acids can be selected with RasMol. They include acidic, basic, aromatic (side chains having ring structures), and aliphatic (side chains without ring structures) amino acids.

- Identify the amino acids in crambin that belong to each family, and place your results in Table 4.1. The commands that must be typed in the Command Line window are shown as follows. Select, color, and restrict one family of amino acids at a time. Display each using the "ball & stick" view from the Display menu. It is not necessary to list the group numbers or to list a particular amino acid more than once per family.

select acidic (or basic, or aromatic, or aliphatic)	selects only acidic (basic, aromatic, or aliphatic) amino acid residues
color orange (or blue, or magenta, or green)	colors acidic (basic, aromatic, or aliphatic) amino acid residues orange (blue, magenta, or green)
restrict acidic (or basic, or aromatic, or aliphatic)	selects only acidic (basic, aromatic, or aliphatic) amino acid residues

TABLE 4.1 Amino Acids of Each Family Present in Crambin		
Amino Acid Family	**Number of Amino Acids in Each Family Present in Crambin**	**Names of the Amino Acids in Each Family Present in Crambin**
Acidic		
Basic		
Aromatic		
Aliphatic		

- To view the entire protein, do the following.

select all	selects all the amino acids

- Choose "ball & stick" from the Display menu.
- Choose "cpk" from the Colors menu.

Part II: Secondary Protein Structure

The following instructions assume that Part II directly follows Part I. If this is a new session, the first command to be typed should be "load crambin.pdb," as in Part I. Colors should be "cpk." When typing commands, be sure to leave a space where a space is indicated.

Two major types of secondary structures can be identified in the crambin protein: the α-helix and the β-sheet.

select helix	selects all the helical regions of crambin
color blue	colors the helices blue

- The α-helix is most easily seen in "ribbons" or "strands" views. Select "strands" if your computer is slow.
- Rotate the molecule by clicking and dragging the mouse or by using the scroll bars.
  1. How many helices do you see? _____
- Rotate the molecule to see how the helices are joined together at one end by a short stretch of amino acids that is at right angles to the helices.
- Select "ball & stick" from the Display menu. The amino acids of the α-helices will remain blue.
- Find amino acids 19 and 22 (19 is at the end of the longer α-helix, and 22 is in the loop between the two helices).
  2. What is the name of the amino acid at these locations? _____
     _____
  3. Describe the side chain of this amino acid. _____
     _____
     _____

select pro	selects only proline residues
restrict pro	shows only proline residues

  4. How many proline residues are found in crambin? _____
  5. How many are part of an α-helix (colored blue)? _____
  6. How many are in the middle of an α-helix? _____
     Proline residues tend not to be found in the middle of α-helices.

select helix	selects all the α-helices of crambin
restrict helix	shows only the α-helices

- Select "ball & stick" from the Display menu. Select "group" from the Colors menu. This colors the residues of the α-helices according to their position in the α-helix and will make it easier to distinguish the other residues in each α-helix. Click on the amino acid residues of the longer α-helix, one at a time.
  7. How many amino acids are in this longer α-helix? _____
  8. Give the name and number of any three amino acids in the longer α-helix. *Alert:* Make sure you have the Command Line window open wide enough or you may not see the entire group number. _____
     _____

select all	selects all the atoms

- Select "ball & stick" from the Display menu.

select sheet	selects all sheet structures of crambin
color yellow	colors the sheets yellow

- Select "ribbons" from the Display menu.

select helix	selects all of the helices of crambin
color blue	colors the helices blue

- Select "ribbons" from the Display menu.
9. How many sheet structures do you see? _____

select helix, sheet	selects both the helices and the sheets
hbonds on	turns on the hydrogen bonds in the helices and sheets

10. Where are the hydrogen bonds in the β-sheets, among the atoms of one strand or between the atoms of parallel strands? _____

11. Where are the hydrogen bonds in the α-helices? _____

select all	selects all the amino acids
hbonds off	turns off the hydrogen bonds

- Select "ball & stick" from the Display menu and "cpk" from the Colors menu.

Part III: Tertiary Protein Structure

The following instructions assume that Part III directly follows Part II. If this is a new session, the first command to be typed should be "load crambin.pdb," as in Part I. Colors should be "cpk." When typing commands, be sure to leave a space where a space is indicated.

The sulfur atom in one cysteine residue can join with the sulfur atom in another cysteine residue to form a disulfide bond. Disulfide bonds contribute to the tertiary structure of proteins.

- Use the Display menu to select "wireframe" structure and the Colors menu to select "cpk" colors.

select cys	selects all the cysteine residues

- Use the Display menu to select "ball & stick."
- Rotate the molecule. Note the position of each cysteine residue.

ssbonds on	makes the disulfide bonds between cysteine residues visible

- Look for the yellow dashed lines showing the disulfide bridges between cysteine residues. You can try coloring the disulfide bonds other colors to make them more visible (the command is "color ssbonds"; then type the name of the color). The cysteine residues and their disulfide bridges can be seen more easily if the rest of the molecule is removed and the cysteine residues are enlarged.

restrict cys	shows only the cysteine residues and disulfide bonds
zoom 150 (or zoom 125)	magnifies the image

- Identify the cysteine residues that are paired by disulfide bonds and identify the sulfur atoms in each disulfide bond. Use the mouse to click on each molecule or atom, and the corresponding name and position number will be displayed in the Command Line window. Enter the position numbers of each cysteine molecule pair in the first column of Table 4.2, and enter the position numbers of the sulfur atoms of the pair in the second column of Table 4.2.

TABLE 4.2 Cysteine Molecule Pairs and Their Sulfur Atoms	
Cysteine Molecule Pairs	**Sulfur Atoms of S–S Bonds**

- Choose one cysteine residue from each pair that is joined by a disulfide bond, and color that one green. An example follows. Do this for each of the three pairs.

select cys40	selects cys40
color green	colors cys40 green

1. Two of the cysteine residues in crambin are joined by a covalent peptide bond rather than by a disulfide bond. Which two are these? _____
    _____

2. Identify the atom (by name and position) in each of these cysteine molecules that is part of the peptide bond. _____
    _____

Part IV: Protein Structure of α-Amylase

Now you will view the porcine α-amylase protein.

zap	removes all the information, including the file
load amypig.pdb	loads the protein data bank file for α-amylase

script amylase.txt	loads the script file (contains information specific to α-amylase protein)
select helix	selects all the helical regions of α-amylase
color blue	colors the helices blue
hbonds on	displays the hydrogen bonds in the helices

- Select "ribbons" from the Display menu; if your computer is slow, choose "strands."

select sheet	selects all the sheet structure regions of α-amylase
color yellow	colors the sheets yellow
hbonds on	displays the hydrogen bonds between sheets
restrict helix, sheet	removes everything but the helices and sheets

- Select "ribbons" from the Display menu. Rotate the molecule so that you see down the barrel of the $(\alpha/\beta)_8$ configuration.

Eight β-strands should be in the center of the barrel, with eight helices around the outside. (Some β-strands remain on the outside of the barrel.)

select active	selects the active site of α-amylase; defined in the amylase.txt file (based on published information)

- Select "ball & stick" from the Display menu.

cpk 200	displays the active-site residues in ball & stick form at size 200
color green	colors the active-site residues green
zoom 150	magnifies the region of interest

- Select "spacefill" from the Display menu.

select all	

- Select "spacefill" from the Display menu. Use the Options menu to turn off the heteroatoms, mostly water, that obscure the view. Just click on "heteroatoms" with the mouse.
- Try other views under the Display menu to help see the active site in relation to the structure of the protein.

Part V: Active-Site Region of α-Amylase

The following instructions assume that Part V directly follows Part IV. If this is a new session, replace the first three commands that follow with the first three commands of Part IV.

reset	resets to normal view
select all	selects everything
hbonds off	turns off the hydrogen bonds

- Select "wireframe" from the Display menu.

One way to study the active site is to use an inhibitor molecule that binds to the enzyme. Since the inhibitor cannot be cleaved, it becomes trapped in the active site. The inhibitor present in the active site of α-amylase is a nonhydrolyzable disaccharide called daf (daf = 1,4-deoxy-4-[5-hydroxymethyl-2,3,4-trihydroxy-cyclohex-5,6-enyl] amino fructose). The nonhydrolyzable disaccharide has two glucose (glc) residues attached to one end and one glc molecule attached to the other end. Thus, the nonhydrolyzable disaccharide occupies the third and fourth positions of the active site.

select glc, daf	selects the inhibitor molecule (which contains glc and daf) blocking the active site

- Select "ball & stick" from the Display menu.

In addition to certain amino acids that are important for enzyme activity, α-amylase contains two ions, Ca^{2+} and Cl^-, which are required for proper function. Because these atoms are difficult to see, they will be enlarged here to help you find them.

cpk 200	sets the inhibitor size at 200
color cyan	colors the inhibitor cyan
select Cl	selects the chloride ion
cpk 300	sets the chloride ion size at 300
color magenta	colors the chloride ion magenta
select Ca	selects the calcium ion
cpk 300	sets the calcium ion size at 300
color red	colors the calcium ion red
zoom 150	magnifies the region of interest
select amino	selects all amino acid residues

- Rotate the image to help see the position of the heteroatoms (i.e., inhibitor, Cl^-, Ca^{2+}, water) in the protein.

restrict hetero	restricts the atoms to heteroatoms
select active	selects the active site

- Select "ball & stick" from the Display menu.

cpk 200	sets the size at 200

- Use the mouse to click on the amino acid groups on the screen; the Command Line will identify each amino acid residue by number.
  1. Identify the 13 amino acid residues at the active site by name and number. *Alert:* Make sure the Command Line window is open wide enough to show the entire group number. _____
  _____
  _____
  _____
  _____

  2. Using Table 4.3, determine how many of the 13 amino acids at the active site are nonpolar, polar, and electrically charged.
  Nonpolar: _____
  Polar: _____
  Charged: _____

TABLE 4.3 Properties of Amino Acid Side Chains	
Properties of Side Chains (R groups)	**Amino Acids**
Nonpolar	gly, ala, val, leu, ile, met, phe, trp, pro
Polar	ser, thr, cys, tyr, asn, gln
Electrically charged	asp, glu, lys, arg, his

- To see the $(\alpha/\beta)_8$ barrel formation along with the cleft in the active site, do the following steps.

restrict dom_a	restricts view to only the $(\alpha/\beta)_8$ barrel formation
color yellow	colors the $(\alpha/\beta)_8$ barrel yellow

- Select "ribbons" from the Display menu. The helices and sheets of the $(\alpha/\beta)_8$ barrel formation are now visible. Rotate the image to see down the middle of the barrel, through the hole in the center.
- Select "spacefill" from the Display menu and notice how the hole in the center disappears.

select active	selects the active-site amino acids
color red	colors the active site red

- Rotate the image until the cleft of the active site is clearly visible. (The active site may be on the opposite side.)

select glc, daf	selects the inhibitor molecule
color green	colors the inhibitor green

- Select "spacefill" from the Display menu and watch as the inhibitor fills the active-site cleft. Rotate the image to see how the inhibitor fits within the active-site region.

Part VI: Disulfide Bridges in α-Amylase

The following instructions assume that Part VI directly follows Part V. If this is a new session, replace the first two steps that follow with the first three steps of Part IV.

reset	resets normal view
restrict none	unselects everything
select all	

- Use the menus to select "wireframe" structure and "cpk" colors.

select cys	selects all the cysteine residues
color yellow	colors the cysteine residues yellow
cpk 150	enlarges the cysteine residues
zoom 150	magnifies

- Rotate the molecule. Note the position of each cysteine residue by clicking on it to see the group number.
  1. Are cysteine residues that are close in the tertiary structure always close together in the primary structure? _____

ssbonds on	makes the disulfide bonds visible
color ssbonds green	colors the disulfide bonds green
zoom 200	magnifies

- Look for the green dashed lines showing the disulfide bridges between cysteine residues. (If green does not show up well on your screen, color the disulfide bonds another color with the command "color ssbonds yellow," for example.)
- The cysteine residues and the disulfide bridges can be seen more easily if you remove the rest of the molecule with the following steps.

restrict cys	shows only cysteine residues and disulfide bonds
color cpk	colors the atoms with conventional colors
zoom 125	magnifies

- *Note:* You may try other values for "zoom" to make the cysteine residues easier to see and count.
- Use the mouse and the Command Line window to identify each cysteine residue by its position number. Choose one cysteine residue from each pair, and color that one yellow. Leave the other one in the pair in cpk colors. An example follows.

select cys115	selects cys 115
color yellow	colors cys 115 yellow

- Continue as in the preceding example for each pair.
  2. How many cysteine residues are in porcine α-amylase? _____
  3. Which atom of the cysteine molecule participates in disulfide bond formation? _____
  4. Are all the cysteine molecules in pairs held by disulfide bonds? _____ If not, which cysteine residues (by number) are not? _____
     _____
  5. Do you think it is possible to predict from just the primary amino acid sequence which cysteine residues will form disulfide bonds? Why?
     _____
     _____

EXERCISES

EXERCISE 4.1
Construct an Image of α-Amylase

To demonstrate your competence using RasMol, perform the following exercise. Reload the α-amylase protein data bank and script files, as you did at the start of Part IV. Then using the RasMol Command Line and your knowledge of the program, construct a view of α-amylase that includes the following elements. Your instructor will view your final image either on the screen or from a color printout.

1. α-helices colored blue in ribbon view
2. β-sheets colored orange in ribbon view
3. active-site amino acid residues colored red in ball & stick view
4. inhibitor molecule colored green in ball & stick view
5. calcium ion colored magenta in spacefilling view
6. chloride ion colored cyan in spacefilling view
7. cysteine residues colored yellow in ball & stick view
8. disulfide bonds on
9. rotate the molecule so that you see down the barrel of the $(\alpha/\beta)_8$ configuration

EXERCISE 4.2
View Information Needed to Generate Images

To appreciate the amount of information needed to view proteins with this program, open the file amypig.pdb in a word processing program or text editor. It should be opened as a text file. This is a large file that contains the information that allows the RasMol program to construct the three-dimensional image of α-amylase. The first part of the document gives information about the molecule and features (e.g., helices, sheets, ssbonds, etc.) of the molecule. The rest of the document gives the coordinates for every atom in the molecule.

Look through the document for information on the regions of the protein where turns occur. Turns are another form of secondary protein structure. A turn causes the polypeptide chain to curve in a different direction. Turns often occur at the ends of helices or sheet structures.

Based on the information in the text file, how many turns are there in α-amylase? _____

LAB 5

Analysis of α-Amylase Proteins

GOAL

The goal of this lab is to separate proteins by SDS-polyacrylamide gel electrophoresis (SDS-PAGE) so that the sizes and immunological reactivity of α-amylases from several sources can be compared.

OBJECTIVES

After completing Lab 5, you will be able to

1. explain the basis of protein migration through SDS-polyacrylamide gels when placed in an electric field
2. explain the need for denaturation and reduction when separating proteins by size
3. estimate the molecular weight of a protein following SDS-PAGE using proteins of known molecular weights as standards
4. explain the process of Western blotting
5. determine whether an unknown sample contains a protein that is immunologically similar to either human or *Bacillus* α-amylases

INTRODUCTION

NATIVE GEL ELECTROPHORESIS

Electrophoresis is a technique that separates charged molecules in an electric field. Negatively charged molecules migrate in an electric field toward the anode (positive electrode); positively charged molecules move toward the cathode (negative electrode). In polyacrylamide gel electrophoresis (PAGE), molecules move through a porous gel matrix, and their rate of migration depends upon the charge, size, and shape of the molecule. Proteins can have a net positive, negative, or no charge at a particular pH because the side chains of many amino acids carry charges. Additionally, the gel matrix acts as a sieve, allowing small molecules to migrate faster than large molecules. Thus, proteins migrate through polyacrylamide gels mainly on the basis of their charge-to-mass ratio. That is, the greater the charge for two proteins of the same mass, the faster the protein migrates; conversely, the larger the size for two proteins of the same charge, the slower the protein migrates. This technique is called native, or nondenaturing, gel electrophoresis and is most commonly used when the native conformation, and hence the activity of the protein, must be maintained.

DENATURING GEL ELECTROPHORESIS

Since proteins of different sizes can have similar charge-to-mass ratios, these proteins would migrate similarly through native polyacrylamide gels and thus would not be separated from each other. Another problem is that many proteins are made of multiple subunits that make the molecule too large to separate easily on native polyacrylamide gels. These problems can largely be overcome by denaturing the proteins prior to electrophoresis by treatment with the negatively charged detergent, sodium dodecyl sulfate (SDS), a reducing agent, and heat. This technique is called *denaturing gel electrophoresis* or *SDS-polyacrylamide gel electrophoresis* (SDS-PAGE). SDS binds strongly to proteins, giving them a negative charge and causing them to lose their three-dimensional structure and unfold into rod-like shapes. The reducing agent (e.g., β-mercaptoethanol) breaks disulfide linkages, further disrupting the tertiary and quaternary structure of proteins. The amount of SDS that binds to proteins is proportional to the size of the protein, giving all proteins the same charge-to-mass ratio. Since all SDS-coated polypeptides have a strong negative charge and a similar rod-like shape, they will migrate to the positive electrode primarily on the basis of molecular weight. Small polypeptides migrate to the bottom of the gel (positive pole) faster than large polypeptides, which are retarded by the gel matrix.

APPLICATIONS OF SDS-PAGE

SDS-PAGE can be used to determine the molecular weight of a protein by comparing its relative distance of migration to that of protein standards whose molecular weights are known. SDS-PAGE can also be used to assess the purity of protein preparations. The presence of a single polypeptide band on a SDS-polyacrylamide gel is one measure of protein purity (although the preparation could still contain contaminants of the same molecular weight). SDS-PAGE can also be used to estimate the relative amounts of a protein by comparing the thickness of polypeptide bands.

A specific protein can be identified after fractionation on an SDS-polyacrylamide gel by exposing all the proteins to a specific antibody. This is generally done after the proteins on the gel have been transferred to a special membrane by blotting. The membrane is then incubated with an antibody that binds to the specific protein. A radioactive, an enzymatic, or a fluorescent tag is used to detect the protein-antibody complex. This method of protein detection is called *immunoblotting,* or *Western blotting.*

GENERAL LABORATORY OVERVIEW

This lab is divided into two sections:

Lab 5A: Different preparations of α-amylase will be resolved on two SDS-polyacrylamide gels. One gel will be stained so that all of the proteins can be visualized and the molecular sizes of the different α-amylases can be estimated.

Lab 5B: The proteins in the second gel will be transferred to a membrane and probed with anti-α-amylase antibodies to identify which proteins are α-amylases.

TIMELINE

Day 1   Lab 5A   Prepare samples (~1 hour)

Day 2   Lab 5A   Load and run gels (~2 hours)
                    Stain one gel with Coomassie Blue (20–30 minutes)

             Lab 5B   Set up transfer of other gel to nitrocellulose (20–30 minutes)
                    Electrophoretic transfer of protein to nitrocellulose (1 hour)

Day 3   Lab 5B   Detect proteins immunologically (~3 hours)

LAB 5A

SDS-PAGE

BACKGROUND

DENATURATION AND REDUCTION OF PROTEINS

Prior to electrophoresis, protein samples are heated in the presence of SDS (to denature the proteins) and a reducing agent such as β-mercaptoethanol (to break disulfide linkages within proteins or between protein subunits). SDS is a negatively charged detergent that binds to the hydrophobic regions of proteins, causing them to unfold into extended polypeptide chains. The amount of SDS that binds is proportional to the size of the protein; approximately one molecule of SDS binds for every two amino acids. The resulting large negative charge conferred by SDS overwhelms any native protein charge and causes all proteins to migrate toward the positive electrode when a voltage is applied. Since the proteins are denatured, their native structure is completely unfolded and all have about the same shape. The end result is that proteins separate primarily according to their size or molecular weight on SDS-PAGE gels. Small proteins migrate to the positive electrode at the bottom of the gel faster than larger proteins.

POLYACRYLAMIDE GELS

Polyacrylamide is a synthetic polymer or chain of acrylamide monomers. These acrylamide chains can be cross-linked to each other by the addition of bisacrylamide during the polymerization reaction. Polymerization occurs in the presence of an initiator (ammonium persulfate) and a catalyst (N,N,N′,N′-tetramethylethylenediamine [TEMED]). The bisacrylamide cross-links cause the chains to form a mesh-like structure in which the holes of the mesh represent the pores that retard protein migration through the gel. At higher acrylamide concentrations, the mesh becomes tighter with smaller pores that more strongly retard the migration of proteins.

Acrylamide concentrations in the range of 10% to 15% separate proteins of about 12,000 to 70,000 daltons. Proteins can be separated on polyacrylamide gels of a single concentration; under these conditions the acrylamide concentration is chosen such that the proteins of interest migrate into and through the gel. Larger proteins are separated on gels with lower concentrations of acrylamide, and vice versa. Sometimes a gradient of acrylamide is used so that the highest concentration, which retards small proteins, is at the bottom of the gel and the lowest concentration, which retards large proteins, is near the top. This means that proteins of widely different sizes can be separated and that proteins of similar sizes may be resolved.

COOMASSIE BLUE STAINING

One way of visualizing the separated protein bands is to stain the gel with Coomassie Blue, a dye that binds strongly to proteins. Coomassie Blue can detect as little as 0.1 μg of protein per band. Before the protein bands can be stained, they must be bound, or fixed, to the gel matrix, and the coating of SDS must be removed so that the polypeptides are accessible to the dye. This is accomplished by soaking the gel in a solution of acetic acid to fix the proteins and isopropanol to remove the SDS. Then the gel is stained with a solution of Coomassie Blue dye. Frequently, the fixing and staining steps are combined and the gel is soaked in a single solution of acetic acid, isopropanol, and dye. After the polypeptides are stained, the gel must be "destained" with a dilute acetic acid solution so the protein bands become visible against a clear background.

LABORATORY OVERVIEW

In this laboratory, you will use SDS-PAGE to analyze the protein composition of several α-amylase-containing samples, i.e., crude lysates of *Bacillus* cells, saliva, and several preparations of commercially purified α-amylases. You will run identical samples on two commercially available precast SDS-polyacrylamide gels (either 10% acrylamide or 4–20% acrylamide). One gel will be stained with Coomassie blue so that all of the polypeptide bands can be visualized (Fig. 5A.1). You will estimate the molecular weight of each purified α-amylase, based on their migration relative to protein molecular weight markers of known sizes. Then you will tentatively identify which band in the crude preparations could possibly be α-amylase, on the basis of size. The second gel will be used for Western blotting (Lab 5B), which will allow you to identify α-amylase in the crude preparations.

SAFETY GUIDELINES

Handle your own saliva samples only. Dispose of any saliva-contaminated pipettes, tips, and tubes in autoclave bags.

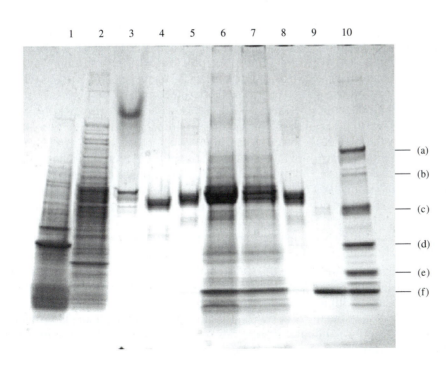

FIGURE 5A.1

Coomassie-stained SDS-polyacrylamide (10%) gel. Lane 1: lysate of *B. licheniformis;* lane 2: lysate of *B. amyloliquefaciens;* lane 3: commercial α-amylase from *B. licheniformis;* lane 4: commercial α-amylase from *A. oryzae;* lane 5: commercial α-amylase from porcine pancreas; lanes 6, 7: saliva; lanes 8, 9: unknown proteins; lane 10: protein molecular weight markers— (*a*) phosphorylase b (97.4 kd), (*b*) bovine serum albumin (66.2 kd), (*c*) ovalbumin (42.7 kd), (*d*) carbonic anhydrase (31 kd), (*e*) trypsin inhibitor (21.5 kd), (*f*) lysozyme (14.4 kd).

Acrylamide, before it polymerizes, is a potent neurotoxin. Always wear gloves when handling polyacrylamide gels.

The current in a gel electrophoresis device is extremely dangerous. Always turn off and disconnect the power supply before removing the lid or touching the gel. Make sure that the counter where the gel is being run is dry.

Wear gloves and handle Coomassie blue solutions carefully; this dye binds to proteins tightly and can stain skin and clothing.

PROCEDURE

This procedure is set up such that each group can work independently to run two gels or two groups can work together to run two gels.

PREPARE THE SAMPLES

☐ 1. Each group will be given a 1.5-ml tube of overnight culture of *Bacillus amyloliquefaciens* cells and a tube of overnight culture of *Bacillus licheniformis* cells. Centrifuge the tubes of cells in a micro-centrifuge at maximum speed for 1 minute to pellet the cells. Pour off the culture medium into a container designated for liquid biological waste.

If two groups are working together, then each group will be given a tube of either *Bacillus amyloliquefaciens* or *Bacillus licheniformis* cells.

☐ 2. Add 50 μl of TE (10 mM Tris-HCl, pH 8, 1 mM EDTA) to each tube and vortex vigorously to resuspend the pellet of cells. Check to see that you have homogeneous cell suspensions. Turn each tube upside down and flick it with your finger to disperse the mixture along the walls of the tube. Look to see if there are any clumps of cells. If there are, continue to vortex until the cell mixture is completely homogeneous.

☐ 3. *Bacillus* cells have to be pulverized with glass beads to be broken open, or lysed, because of their thick cell wall. The resulting lysate contains the contents of the cells, as well as broken cell membranes. Label new 1.5-ml tubes as "*B. amyloliquefaciens* lysate" or "*B. licheniformis* lysate." Add 0.1 ml of glass beads to each tube. To do this, use a 1.0-ml serological pipette and a pipette pump to draw approximately 0.5 ml of the glass bead suspension into the pipette. Allow the glass beads to settle to the bottom of the pipette. With the pipette tip still in the bead container, slowly dispense the beads until the level of the beads aligns with a major marking (e.g., 0.7 or 0.8 ml). Now, position the pipette inside one of the newly labeled microcentrifuge tubes, with the tip near the bottom of the tube, and dispense 0.1 ml of beads. It is important to add the beads to the bottom of the tube; do not place any of the glass beads on the edge of the tube or on the lid as this may prevent the tube from closing properly.

If two groups are working together, then each group needs to prepare only one tube.

☐ 4. Using a micropipette, transfer all of the resuspended *Bacillus* cells from each of the original tubes (step 2) to the appropriately labeled tube with the glass beads. Cap the tubes tightly.

☐ 5. Vortex the tubes at maximum speed for 1 minute; then place them on ice for 1 minute. Repeat this sequence four more times for a total vortexing time of 5 minutes.

6. Let the tube of lysed cells sit on ice for a minute or so to allow the beads to collect at the bottom. Using a micropipette, transfer 30 μl of each *Bacillus* lysate to a new, labeled 1.5-ml tube. Add 30 μl of 2× SDS-PAGE Loading Buffer (120 mM Tris-HCl, 4% SDS, 1% β-mercaptoethanol, 20% glycerol, 0.01% Bromophenol Blue) to each tube. (The glycerol increases the density of the sample so it can be loaded on the gel, and the blue dye permits tracking of the sample during electrophoresis.) Vortex to mix.

7. Each member of the group will collect some saliva by placing a disposable plastic bulb-type transfer pipette under her or his tongue and gently squeezing the bulb to draw saliva into the pipette shaft. Dispense the saliva into a 1.5-ml tube. *HANDLE ONLY YOUR OWN SALIVA SAMPLE.* Use a scissors to cut off the tip of a plastic micropipette tip and transfer 30 μl of saliva to a new, labeled 1.5-ml microcentrifuge tube (pipette slowly because the saliva is very viscous). Add 30 μl of 2× SDS-PAGE Loading Buffer to the new tube and vortex to mix.

   If two groups are working together, one member of each group will be a saliva donor.

8. Add 30 μl of each of the following samples to new, appropriately labeled 1.5-ml tubes:

   commercial α-amylase from *Bacillus licheniformis*
   commercial α-amylase from *Aspergillus oryzae*
   commercial α-amylase from porcine pancreas
   unknown #1
   unknown #2

   If two groups are working together, then only one set of the previously listed samples needs to be prepared. The two groups should coordinate their efforts.

9. Add 30 μl of 2× SDS-PAGE Loading Buffer to each of these tubes. Vortex to mix.

10. Transfer 30 μl of the diluted protein molecular weight markers to a new, labeled tube. The protein molecular weight markers have been already mixed with the SDS-PAGE Loading Buffer and are ready for heating. This is a mixture of proteins of known sizes whose relative distances of migration will be used to generate a standard curve from which you will determine the molecular weights of your proteins of interest. Your instructor will provide the names and molecular weights of the protein standards you are using.

11. Heat all 10 tubes for 5 minutes at 95°C.

12. Microcentrifuge the tubes of *Bacillus* lysates for 2 minutes to pellet the cell debris. Using a micropipette, transfer each supernatant solution to a new, labeled tube.

13. If the gels are to be run during another lab session, store the samples in the freezer.

LOAD AND RUN THE GELS

You will load and run two gels; each will be loaded in the same way with the 10 samples you prepared in the preceding section.

1. *WEAR GLOVES WHEN HANDLING POLYACRYLAMIDE GELS.* Remove each gel from its package and rinse the outside of the cassette under running water. Follow the directions provided by

the manufacturer of the gel for removing or inserting the combs. Measure and record in your notebook the dimensions (length $\times$ width, in centimeters) of a gel (you can include the wells in your measurement). You need this information for the electrophoretic transfer in Lab 5B.

☐ 2. Assemble the gel cassettes in the electrophoresis device(s) according to the manufacturer's directions. The longer plate generally faces outward, and the shorter, notched plate faces the buffer chamber. Add SDS-PAGE Running Buffer (25 mM Tris, 192 mM glycine, 0.1% SDS) to the upper and lower buffer chambers.

☐ 3. Remove any debris and air bubbles from each well by gently flushing the wells with Running Buffer using a Pasteur pipette.

☐ 4. If the samples have been frozen, heat the tubes for 5 minutes at 95°C. Briefly spin the tubes in a microcentrifuge. Place all tubes on ice.

☐ 5. Use a micropipette to load 10 μl of each sample in the order listed as follows. (If the gels have 12 lanes, load 10 μl of 1$\times$ SDS-PAGE Loading Buffer into each of the two outside lanes.)

Lane #	Contents
1	*Bacillus amyloliquefaciens* lysate
2	*Bacillus licheniformis* lysate
3	commercial α-amylase from *Bacillus licheniformis*
4	commercial α-amylase from *Aspergillus oryzae*
5	commercial α-amylase from porcine pancreas
6	saliva A
7	saliva B
8	unknown #1
9	unknown #2
10	protein molecular weight markers

☐ 6. Flush the wells of the second gel with Running Buffer just before loading. Load this gel in the same manner as the first gel.

☐ 7. Run the gels at about 125 volts until the Bromophenol Blue tracking dye migrates to 0.5–1 cm from the bottom of the gel (this generally takes less than 1 hour).

☐ 8. While the gels are running, prepare the absorbent paper and nitrocellulose membrane for the electrophoretic transfer of Lab 5B.

☐ 9. When the gels are finished running, turn off and disconnect the power supply, remove the lid of the electrophoresis device, discard the running buffer, and remove the gel cassette(s).

STAIN ONE GEL

☐ 1. ***WEAR GLOVES WHEN HANDLING POLYACRYLAMIDE GELS AND COOMASSIE BLUE SOLUTIONS.*** Add about 100 ml of Coomassie Blue Stain Solution (25% isopropanol, 10% acetic acid, 0.05% Coomassie Blue) to a shallow plastic container (e.g., a Rubbermaid sandwich container with a lid). Since this container will be placed in a microwave, there must be a small release hole punched in the lid.

☐ 2. Open the gel cassette by gently prying the plates apart with a small metal spatula. The gel will adhere to one plate—pay attention to which plate it adheres. Use the spatula to cut off the upper right corner of the gel, near the last lane (lane 10). If the gel sticks to the notched, back plate, the upper right corner of the gel corresponds to the upper right corner of the plate. If the gel sticks to the longer front plate, then the orientation will be reversed. Submerge the plate, gel-side down, into the stain solution and gently pry the gel off the plate by inserting the spatula under one corner of the gel. Place the lid securely on the container.

☐ 3. Microwave for 2 minutes or until the solution is boiling.

☐ 4. Rock the gel and the hot solution for 5 minutes at room temperature.

☐ 5. Discard the staining solution and rinse the gel briefly with 100–200 ml of dH$_2$O.

☐ 6. Add 100 ml of Coomassie Destain Solution (10% acetic acid). Place a Kimwipe in the container to absorb the dye. Replace the lid.

☐ 7. Microwave for about 1 minute (the tracking dye will disappear if the solution is microwaved too long). Rock the gel and the hot Destain Solution at room temperature for 15 minutes. Protein bands should be visible after this period of time. You can reduce background staining by replacing the Destain Solution with fresh Destain Solution. The gel can be stored until the next lab session (or indefinitely) in Destain Solution.

If you do not have a microwave, you can stain the gel by incubating it overnight at room temperature in the Coomassie Blue Stain Solution. Replace with Destain Solution and allow to incubate overnight or until protein bands are visible. You can speed up the destaining process by changing the Destain Solution several times.

☐ 8. Place the stained gel on plastic wrap and make photocopies of it for all members of the group or groups. Alternatively, you can take a photograph of the gel, but a direct photocopy of the gel works best for the analysis following Lab 5B because the image is not reduced in size.

DATA ANALYSIS

- Label each lane (i.e., identify the contents) on the photocopy or photograph of your gel. Each of the lanes loaded with the purified α-amylases (lanes 3–5) should have one predominant band that is α-amylase. The lanes loaded with the *Bacillus* lysates and saliva samples should have many bands.

- Identify which band in lanes 1 through 7 could possibly be an α-amylase. To aid in your denotation, make an overlay of the gel by placing a piece of plastic wrap or a sheet of overhead transparency film over the gel (or its photocopy) and marking the wells and the edges of the gel. Use a fine-tipped marking pen to mark on the overlay the location of the molecular size standards. Mark the band in each of the commercial preparations (lanes 3–5) that is most likely α-amylase. Mark on your overlay which band in the saliva samples (lanes 6–7) is most likely α-amylase. Explain your reasoning. Mark on your overlay which band in the *Bacillus* lysates (lanes 1–2) could be α-amylase. Explain your reasoning. Include this overlay in your notebook, along with the photocopy of your gel.

- Next, estimate the molecular sizes of the different α-amylases and your unknown proteins. For this, you first must construct a standard curve that plots relative mobility versus molecular weight. Relative mobility (R_f) is calculated by dividing the distance a protein band migrates by the distance that the tracking dye migrates. The tracking dye (Bromophenol Blue) is small and migrates faster than proteins; its migration approximates the movement of ions in the electrophoretic front. The relative mobility of a protein is more accurate than the absolute distance it migrated because differences between lanes are eliminated (notice that the tracking dye migrated slower in the outside lanes than in the center lanes).

- Measure the distance that each protein molecular weight marker migrated (measure from the bottom of the well to the middle of each band), and measure the distance the tracking dye migrated in lane 10. (You can measure from your photocopy of the gel or directly from the gel.) Calculate the relative mobility (R_f) of each. Organize your data into a table in your notebook. Generate a standard curve by plotting the R_f (x-axis) for each protein molecular weight marker versus its molecular weight in daltons (y-axis) on semilog graph paper. Draw a straight best-fit line through the data points that are on the linear portion of the curve. (It is possible that the largest and/or smallest molecular size standards do not fall in the linear range, resulting in an S-shaped curve. If this is the case, then draw your line through at least four points that clearly lie on the linear portion of the curve.) Include your standard curve in your notebook.

- Calculate the R_f of the putative α-amylases (lanes 1–7) and the unknown proteins (lanes 8–9) by measuring the distance each migrated (measure from the bottom of the well to the middle of each band) and the distance the tracking dye migrated in each lane. Add these data to your table. Using your standard curve, estimate the molecular weight of each protein band you identified as α-amylase in lanes 1–7. (Does your estimation of the molecular weight of porcine α-amylase agree with what you determined in Lab 4?) Calculate the molecular weights of your two unknown proteins in lanes 8 and 9. You may be able to identify one or both of your unknown proteins on the basis of size (one is an α-amylase and the other is one of your protein molecular weight markers). You will be able to identify both at the end of Lab 5B.

QUESTIONS

1. Explain why polypeptides separate mainly on the basis of size or molecular weight during SDS-PAGE.
2. Suppose you ran a 15% SDS-polyacrylamide gel and the proteins of interest were still packed together near the top of the gel. You have enough of your samples to run another gel. Should this new gel contain a higher percentage or a lower percentage of polyacrylamide? Why?
3. In Lab 2, you concluded that the specific activity of a commercial preparation of α-amylase purified from *Bacillus licheniformis* was greater than that of α-amylase in saliva. Explain whether the results of your SDS-polyacrylamide gel support this conclusion.
4. Compare the relative amounts of α-amylase in the lysates of *Bacillus* cells and human saliva.
5. List three types of information provided by SDS-PAGE.

LAB 5B

Western Blotting

BACKGROUND

Immunoblotting, or Western blotting, is a technique that detects proteins, separated by SDS-PAGE, by their reaction with a specific antibody. Before interaction with specific antibodies, the separated proteins are transferred or blotted from the gel to a nitrocellulose or nylon membrane that binds proteins nonspecifically. After transfer, the remaining binding sites on the membrane are blocked with a mixture of unrelated proteins to eliminate any further reactions with the membrane. The membrane is then incubated with an antibody to the protein of interest (this antibody is called the primary antibody). Since all the protein-binding sites on the membrane are blocked, the primary antibody can bind to the membrane only if it interacts with its specific protein. The membrane is washed extensively to remove unbound primary antibodies. While the antigen-antibody complex could be detected directly by virtue of a tag on the primary antibody (i.e., a radioactive isotope, a fluorescent dye, etc.), it is generally detected indirectly because this increases the strength of the signal. For this, the membrane is incubated with another antibody (called a secondary antibody) that is attached to an enzyme. The secondary antibody recognizes and binds to the primary antibody. The antigen-primary antibody-secondary antibody-enzyme complex is detected when the enzyme converts a soluble, colorless substrate into an insoluble, colored product. Thus, colored bands appear on the white membrane wherever a protein interacts with the primary antibody (Fig. 5B.1).

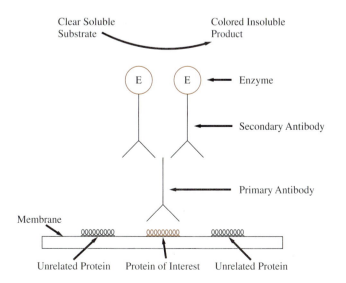

FIGURE 5B.1

Immunological detection of the protein of interest. The view is a cross section of the membrane after transfer of the proteins from the acrylamide gel. The left side of the membrane represents the top of the gel; the right side represents the bottom of the gel.

LABORATORY OVERVIEW

The different α-amylase proteins you resolved in Lab 5A have different molecular weights, and they have different primary sequences of amino acids. The location of the different types of amino acids (i.e., acidic, basic, etc.) throughout each sequence is similar, however, since each folds into the $(\alpha/\beta)_8$ barrel conformation that is characteristic of α-amylase proteins. The amino acid sequence of porcine α-amylase is similar to those of other mammalian α-amylases, but differs from those of bacterial and fungal α-amylases. (See the results of the Basic BLAST Search exercise in Appendix I.) Thus, in this lab, you will use a mixture of primary antibodies made against two different α-amylase proteins so that you will be able to detect a greater number of α-amylase proteins on your Western blot.

These primary antibodies were produced when purified α-amylase protein from a single source was injected into a rabbit. The rabbit's immune system mounted an immunological response to the foreign protein, or antigen, by producing a heterogeneous mixture of antibodies against it. The different antibodies recognize different parts of the antigen molecule; hence, they are termed *polyclonal*. You will use a mixture of polyclonal antihuman salivary α-amylase antibodies and anti–*Bacillus amyloliquefaciens* α-amylase antibodies. Since mammalian α-amylase proteins share regions of conserved amino acid sequence, antibodies made against α-amylase from humans should also recognize α-amylase from pigs. Likewise, antibodies made against α-amylase from *Bacillus amyloliquefaciens* should also recognize α-amylase from *Bacillus licheniformis*.

The α-amylase-antibody complexes will be recognized by antirabbit secondary antibodies, which were produced in goats that were inoculated with rabbit antibodies. The secondary antibodies are coupled to the enzyme, alkaline phosphate (AP), which forms a dark blue-purple precipitate when it reacts with the soluble, colorless substrate 5-bromo-4-chloro-3-indolyl phosphate (BCIP) in combination with nitroblue tetrazolium (NBT).

In Lab 5A, you were unable to identify conclusively which of the many bands in samples of *Bacillus* cells and human saliva was an α-amylase. In this lab, you will be able to identify the α-amylase bands by virtue of the strong and specific interactions between an antibody and its antigen.

SAFETY GUIDELINES

Acrylamide, before it polymerizes, is a potent neurotoxin. Always wear gloves when handling polyacrylamide gels.

Gloves should also be worn when handling nitrocellulose or nylon membranes since the oils and proteins on your bare hands can interfere with the binding of your proteins of interest.

PROCEDURE

TRANSFER PROTEINS FROM THE GEL TO A MEMBRANE

☐ 1. ***WEAR GLOVES WHEN HANDLING NITROCELLULOSE MEMBRANES.*** Cut six sheets of absorbent paper and one sheet of nitrocellulose membrane to the exact size of the gel. (Use a pencil, not a pen, to make cutting lines on the membrane.) It is important that the absorbent paper and membrane are cut no larger than the gel because if they hang over the sides of the gel and touch the electrode plates, the transfer process will be short circuited, resulting in little transfer of protein to the nitrocellulose membrane.

Note: In the package, nitrocellulose membrane is often sandwiched between two pieces of protective paper.

2. Soak the nitrocellulose membrane in ~20 ml of dH$_2$O. Do this by carefully laying the nitrocellulose on the surface of water in a shallow container. Allow the membrane to become wet by capillary action from the bottom for several minutes, then submerge the nitrocellulose membrane (i.e., press it down below the surface of the water with gloved hands) and allow it to soak for several more minutes.

3. Cover the bottom of a shallow container with 20–25 ml of Western Transfer Buffer (48 mM Tris, 39 mM glycine, 0.037% SDS, 20% methanol). Place the six sheets of absorbent paper in the container and let them soak for several minutes.

4. Put a piece of plastic wrap on the workbench and stack two pieces of the wet absorbent paper on the plastic wrap. (To keep the plastic wrap in place, put a drop or two of water between the plastic wrap and the lab bench.) Gently pry apart the gel plates with a small metal spatula; the gel will adhere to one plate. Cut off the upper right corner of the gel, as described in Lab 5A (it is not necessary to cut off the wells). Place another piece of wet absorbent paper on the gel adhering to the plate. Carefully invert the plate and gently release the gel and paper together onto the stack of papers on the bench by inserting the spatula under one corner of the gel. The gel should now be on top of the three pieces of absorbent paper on the bench. Use the metal spatula to carefully reposition the gel and papers so that all of the edges are aligned.

5. Place the wet nitrocellulose membrane on the gel, aligning the edges. Place the last three pieces of wet absorbent paper on the stack, aligning the edges. Now the gel and nitrocellulose are sandwiched between sheets of absorbent paper. This transfer sandwich will be positioned in the electroblotter with the nitrocellulose membrane between the gel and the positive electrode. This will allow the negatively charged, SDS-coated proteins to move from the gel toward the positive electrode and onto the nitrocellulose membrane (Fig. 5B.2).

6. Remove all air bubbles between the layers of the sandwich by rolling a glass pipette from the center to the edges of the stack. Place the transfer sandwich (without the plastic wrap) on the bottom of the blotting unit. Check the instructions that came with your electroblotter to determine whether the unit blots upward or downward. Arrange the transfer sandwich so that the nitrocellulose membrane is closest to the positive electrode. Wipe up any buffer that surrounds the transfer sandwich to decrease the chances of short-circuiting the transfer.

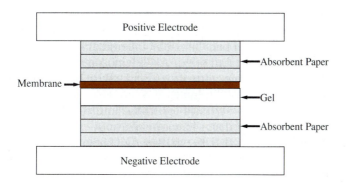

FIGURE 5B.2
Setup for the electrophoretic transfer of proteins from a gel to a membrane.

7. The entire bottom surface of the blotting unit can be used to set up as many transfer sandwiches as will fit. (Two sandwiches can also be stacked for transfer at the same time, as long as dialysis membrane is placed between the absorbent papers of the two stacks. The dialysis membrane should first be soaked in Western Transfer Buffer.)

8. Carefully place the upper electrode on top of the transfer sandwiches. Connect the electrodes and set the power supply so that the current is constant. The current setting should be 0.8 mA per cm^2 of gel surface. For example, if a gel is 9 cm × 8 cm, the current would be set at 58 mA per sandwich:

$$9 \text{ cm} \times 8 \text{ cm} \times 0.8 \text{ mA/cm}^2 = 58 \text{ mA}$$

Increase the current accordingly for each transfer sandwich. Turn on the power supply and transfer for 1 hour.

9. When the transfer is finished, turn off and disconnect the power supply. You or your instructor will disassemble your transfer sandwich. Wearing gloves, discard the absorbent paper and gel. Allow the membrane to air-dry. Write your group number or initials (in pencil) along the upper edge of the membrane, on the side of the membrane that was facing the gel (i.e., the protein-containing side of the membrane). Place the nitrocellulose into a shallow, plastic container or wrap in foil and store until the next lab period.

DETECT PROTEINS IMMUNOLOGICALLY

1. ***WEAR GLOVES WHEN HANDLING NITROCELLULOSE MEMBRANES BECAUSE THE OILS AND PROTEINS FROM YOUR SKIN CAN INTERFERE WITH PROTEIN BINDING.*** If the membrane is dry, rewet it in ~25 ml of 1× Tris Buffered Saline (TBS) (25 mM Tris-HCl, pH 7.5, 137 mM NaCl, 2.7 mM KCl). Wet it by floating it on the surface of the 1× TBS in a shallow container. Once the membrane is wetted, submerge it and let it soak for several minutes.

2. Discard the TBS. Add 30 ml of Western Blocking Buffer (5% dry milk powder, 1× TBS, 0.05% Tween-20) and rock gently at room temperature for 30 minutes. (The milk proteins bind to the remaining protein binding sites on the nitrocellulose membrane. Without this step, the antibodies would bind to these sites, producing a high background.)

3. Discard the blocking buffer and add 30 ml of Primary Antibody Solution (rabbit antibodies against α-amylases from human saliva and *Bacillus amyloliquefaciens* diluted in Western Blocking Buffer). Rock the membrane and antibody mixture gently for 45–60 minutes at room temperature.

4. Discard the Primary Antibody Solution. Briefly rinse the membrane with about 50 ml of TBST (1× TBS, 0.05% Tween-20). Swirl the TBST solution over the membrane several times and discard the solution.

5. Wash the membrane further by adding about 50 ml of TBST and rocking gently for 2 minutes at room temperature. Repeat this washing step three more times for a total of four washes of about 2 minutes each. Discard all the wash solutions.

6. Add 30 ml of Secondary Antibody Solution (AP-conjugated goat anti-rabbit antibodies diluted in Western Blocking Buffer). Rock gently for about 45 minutes at room temperature.

7. Discard the Secondary Antibody Solution and briefly rinse the membrane with about 50 ml of TBST (as in step 4).

8. Further wash the membrane by adding about 50 ml of TBST and rocking gently for 2 minutes at room temperature. Repeat washing three more times for a total of four washes. Discard all the wash solutions.

9. Add about 50 ml of AP Reaction Buffer (100 mM Tris-HCl, pH 9.5, 100 mM NaCl, 50 mM $MgCl_2$) to the membrane. Alkaline phosphatase functions best at high (alkaline) pH in a phosphate-free buffer and requires Mg^{2+}. Gently rock for 2 minutes at room temperature.

10. Discard the AP Reaction Buffer. Transfer the membrane to a clean, shallow dish or to a new zippered bag. Make sure that the protein-containing side of the membrane is facing upward. Add 20 ml of AP Substrate Solution (BCIP and NBT in AP Reaction Buffer), and allow the reaction to proceed in the dark or in low light. Alkaline phosphatase catalyzes the removal of a phosphate group from BCIP, and the resulting product is oxidized by NBT to form a dark blue-purple precipitate. Check the membrane occasionally until blue-purple bands appear.

11. When the desired color is obtained, discard the AP Substrate Solution, and add 50 ml of AP Stop Buffer (20 mM EDTA in phosphate buffered saline at pH 7.5) to stop the color reaction. Alkaline phosphatase does not function well at neutral pH; EDTA binds, or chelates, the Mg^{2+}; and the excess phosphate molecules compete with the substrate BCIP. Gently rock for 2 minutes.

12. Discard the AP Stop Buffer and allow the membrane to dry. Make photocopies of it for all members of the group or groups to add to their notebooks. Wrap the membrane in aluminum foil to keep light from degrading the color of the bands.

DATA ANALYSIS

- Identify and number the lanes on your membrane (or its photocopy) to aid in your analysis. The molecular size standards will not be visible (unless you ran prestained or colored protein size markers), and your blot could be a mirror image of your gel; it depends upon which side of the gel was placed against the nitrocellulose membrane. Determine which lanes contain protein bands that reacted with the antibodies. Most lanes will contain one predominant band, although some lanes may have additional faint bands.

- Determine which of the commercial α-amylases (lanes 3–5) reacted with the antibodies. Place your overlay from Lab 5A on your membrane (or its photocopy); you should be able to superimpose the bands of some of the commercial α-amylases on the overlay with colored bands on your immunoblot. Comment on whether or not the bands you identified tentatively as α-amylase in the saliva samples (lanes 6–7) in Lab 5A match those detected by immunoblotting. (Ignore the low molecular weight band in the saliva samples that cross-reacts with the antibodies.) Comment on whether or not the bands you identified tentatively as α-amylase in the lysates of *Bacillus* cells (lanes 1–2) in Lab 5A match those detected by immunoblotting.

- Note whether either of your unknown samples (lanes 8–9) reacted with the antibodies. Use this information and your information on their molecular weights from Lab 5A to identify your unknown proteins. (One is an α-amylase, and the other is one of your protein molecular weight markers.) Justify your reasoning.

QUESTIONS

1. What is the purpose of immunoblotting? What information does it provide?
2. Did all of the commercial preparations of α-amylase react with the antibodies? If not, provide a possible reason.
3. What is the limiting factor in Western blotting? (*Hint:* Think about your answer to question 2.)
4. Lanes 1 and 2 of your Western blot contained the lysates of *Bacillus* cells, and in each lane, the anti-*Bacillus* α-amylase antibodies should have detected a band about the same size as that in lane 3, which contained the commercially purified *B. licheniformis* α-amylase. What could you conclude about the preparation of the lysates if there were no bands of the correct size on your Western blot, but rather only lower molecular weight bands?
5. Write a paragraph to explain the process of Western blotting.

LAB 6

Analysis of DNA Structure Using RasMol

GOAL

The goal of this laboratory is to examine the structure of DNA using the computer program RasMol.

OBJECTIVES

After completing Lab 6, you will be able to

1. identify the nucleotides of DNA as purine or pyrimidine
2. describe the complementary and antiparallel nature of two strands of DNA
3. describe the formation of phosphodiester bonds
4. identify the major and minor grooves on a DNA molecule
5. describe one type of protein-DNA interaction

BACKGROUND

The hereditary material of most organisms is made of DNA, or deoxyribonucleic acid, which is a polymer of nucleotides. A nucleotide comprises a purine base (adenine, guanine) or a pyrimidine (cytosine, thymine, uracil) base linked to a sugar residue (ribose or deoxyribose) on which one or more phosphate groups are attached (Fig. 6.1). In DNA, the deoxynucleotides, adenosine, thymidine, guanosine, and cytidine, are joined together by covalent bonds, called *phosphodiester bonds.* This is a phosphate ester linkage between the 5′-phosphate group on the sugar residue of one nucleotide and the 3′-hydroxyl group on the sugar residue of the next nucleotide (Fig. 6.2). DNA exists as a double-stranded molecule in which the bases of one strand are hydrogen bonded, or base paired, with the bases of the other strand. This base pairing is complementary; adenine (A) always pairs with thymine (T), and guanine (G) always pairs with cytosine (C). The A-T base pair has two hydrogen bonds, and the G-C base pair has three hydrogen bonds (Fig. 6.3). The two strands of DNA form a helical structure, the double helix described by Watson and Crick nearly 50 years ago. The two strands of the double helix are organized in opposite orientations, such that the 5′ end of one strand is aligned with the 3′ end of the other strand. RNA, or ribonucleic acid, is similar in structure

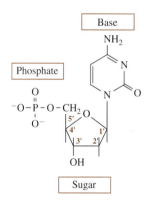

FIGURE 6.1

Nucleotide structure. Deoxycytidine monophosphate is shown. Note the absence of the 2′ OH in the deoxyribose, which is characteristic of nucleotides in DNA.

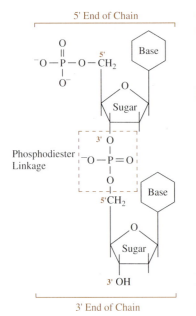

5' End of Chain

3' End of Chain

FIGURE 6.2
Phosphodiester linkage.

Phosphodiester Linkage

to DNA, except ribose is the sugar moiety, the nucleotide uridine is used rather than thymidine, and RNA is single stranded.

BACKGROUND QUESTIONS

1. What is a nucleotide?
2. How does the structure of purine bases differ from that of pyrimidine bases?
3. How does the base pairing between G and C differ from the pairing between A and T?
4. Explain why the two strands of DNA are complementary.
5. Explain the antiparallel nature of double-stranded DNA.

LABORATORY OVERVIEW

In this laboratory, the structure of DNA will be explored using the computer program RasMol. You will use many of the same commands and operations you used to view protein structure in Lab 4. The pdb files needed for this lab can be downloaded from this manual's website at <http://www.mhhe.com/thiel>. (Refer to the Bioinformatics section of Appendix I for information regarding other DNA manipulation programs. Also, the Chime and Protein Explorer programs mentioned in Lab 4 can be used to visualize DNA structure.)

In Parts I through IV, you will explore nucleotide structure, phosphodiester linkages, and the complementary and antiparallel nature of double-stranded DNA. In Part V, you will examine interactions between DNA and a DNA-binding protein.

TIMELINE

This computer exercise takes 1–2 hours; it can be done in class or assigned as homework.

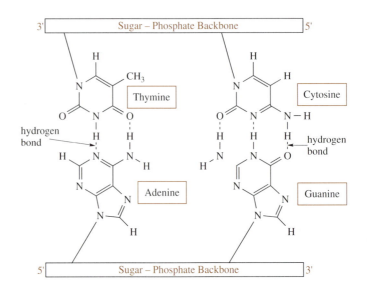

FIGURE 6.3
Base pairing in DNA. Note that there are three hydrogen bonds in the G-C pair and two hydrogen bonds in the A-T pair.

PROCEDURE

USING A MACINTOSH TO VIEW DNA WITH RASMOL

1. The RasMac icon has three circles arranged in a triangle. Follow directions given by the instructor for opening the program from the desktop.
2. There should be two windows: one with a black background titled "RasMol"; the other with a white background titled "RasMol Command Line." Adjust the size of the windows by clicking on and dragging (up, down, left, right) the small box on the lower right-hand corner of each window. The white Command Line window should cover about one-third of the left of the screen, and the black RasMol window should cover about two-thirds of the right side of the screen.
3. Go to step 4 of the PC instructions that follow.

USING A PC TO VIEW DNA WITH RASMOL

1. From the Start Menu, open RasMol. The black "RasMol" window with the menu items above it will be displayed.
2. On the Task Bar at the bottom of the screen, "RasMol Command" will be displayed. Open this by clicking on it. This will open a second white window with a prompt line that accepts typed commands.
3. Adjust the size of the two windows so that both are seen at the same time. You may need to move the Command Line window slightly to the left in order to reduce it from the right side. Move the mouse to the outside line of the window that you want to change until you see a double-ended arrow. Click and hold the mouse while you drag the side of the window to the size desired. The white Command Line window should cover about one-third of the left of the screen, and the black RasMol window should cover about two-thirds of the right side of the screen.
4. In the instructions that follow, the left side of the instruction table presents the commands that should be typed in the Command Line window. The right side of the table gives comments on (explains) the purpose of each command. Type only the command, not the comment. After typing the command, push the "Enter" key on the keyboard. *TO MINIMIZE ERRORS, TYPE COMMANDS CAREFULLY, AND MAKE SURE SPACES ARE PLACED ONLY WHERE SHOWN IN THE COMMANDS. BULLETED ITEMS DIRECT YOU TO USE THE MOUSE TO CHOOSE MENU ITEMS OR MANIPULATE THE IMAGE.*

Part I: DNA Structure

- Open the file "dna1.pdb." Turn the DNA molecule to view it from the side rather than from the end. Use the Display and Colors menu items to view the molecule in different forms and colors.
- Select "ball & stick" from the Display menu and "cpk" from the Colors menu. (In cpk colors, red = O, blue = N, gray = C, and yellow = P.) Turn off the heteroatoms by clicking "heteroatoms" under the Options menu.

Commands—type text below and press "enter"	Comments
select a	selects all of the deoxyadenosine residues
color blue	colors them blue
select g	selects all of the deoxyguanosine residues
color green	colors them green
select t	selects all of the deoxythymidine residues
color yellow	colors them yellow
select c	selects all of the deoxycytidine residues
color red	colors them red

1. How many A residues are there in this strand of DNA? _____
2. How many G residues are there in this strand of DNA? _____
3. How many T residues are there in this strand of DNA? _____
4. How many C residues are there in this strand of DNA? _____

select nucleic	identifies the molecule as nucleic acid for the next step
select backbone	identifies the sugar-phosphate backbone

- Display as "ribbons."
5. How many sugar-phosphate backbones are there in DNA? _____

Recall that the backbones contain alternating phosphate and sugar residues linked with phosphodiester bonds.

- Select "ball & stick" from the Display menu and "cpk" from the Colors menu.

restrict 1,7,18,24	identifies two pairs in the DNA helix
zoom 150	magnifies the image
hbonds on	shows the hydrogen bonds for each pair

6. Identify the two nucleotides in each pair by letter and number. _____
_____

- If you cannot remember which color represents which base, click on the base with the mouse, and the letter and number of the molecule (e.g., G24) will appear in the command line. It is also possible to identify each atom in the molecule by clicking directly on the atom. This may be easier to see if you zoom to 200.

reset	clears everything
select all	selects everything

- View the molecule as "sticks" from the Display menu.

hbonds on	turns on all of the hydrogen bonds
zoom 150	magnifies the image

- Rotate the molecule to see the hydrogen bonds.
  7. Is there a consistent number of hydrogen bonds between certain pairs of nucleotides? _____ If so, what general pattern is seen? _____
  _____

- View using "chain" under the Colors menu.
- Find and identify the major and minor grooves. These appear as hollows or curves where there are no molecules. If you trace each hollow around the three-dimensional structure, you will find that both the large hollow (the major groove) and the smaller hollow (the minor groove) trace a helical path. It may help to change the display to "spacefill" to help see the grooves. Rotate the molecule until they are visible. Turn off the heteroatoms under the Options menu (you may need to turn them on and then off to get rid of them).

zap	closes the file

Part II: Nucleotide Structure

- Open the file "dna1.pdb." Display it in "ball & stick" with "cpk" colors and the heteroatoms off (under the Options menu). (In cpk colors, red = O, blue = N, gray = C, and yellow = P.)

First you will examine the chemical structure of adenosine.

restrict a22	shows only nucleotide A22
zoom 200	magnifies the image

  1. Identify the purine base that contains two nitrogenous rings. Draw a picture of this purine molecule (adenine).

- Rotate the nucleotide to clearly view the 5-carbon deoxyribose ring.
  2. Draw a picture of the deoxyribose ring and label each carbon (1′–5′).

3. Adenine is attached to which carbon in the deoxyribose ring? _____
   _____
4. Find the 3′ carbon of the deoxyribose ring. Does it have an oxygen attached? _____
5. Find the 2′ carbon of the deoxyribose ring. Does it have an oxygen attached? _____
6. Find the 5′ carbon of the deoxyribose ring. What chemical group is attached? _____

Now you will examine the chemical structure of thymidine.

restrict t3	restricts view to nucleotide T3

- To see the nucleotide, choose "ball & stick" from the Display menu. (Keep "cpk" colors and the heteroatoms off.)
  7. Identify the pyrimidine base that contains one nitrogenous ring. Draw a picture of this pyrimidine base (thymine).

select all	

- Display in "ball & stick." Turn off the heteroatoms.

restrict 1,7,18,24	identifies two nucleotide pairs in the DNA helix
hbonds on	shows the hydrogen bonds for each pair

8. Which two nucleotides are purines? _____

_____

9. Which two nucleotides are pyridimines? _____

_____

10. What can you say about nucleotide pairs with regard to the purine-pyrimidine content? _____

_____

zap	closes the file

Part III: Phosphodiester Bond Formation

- Reopen the file "dna1.pdb." Display in "ball & stick" with "cpk" colors and heteroatoms off. (In cpk colors, red = O, blue = N, gray = C, and yellow = P.)

restrict 2	shows only nucleotide 2
select 3	selects nucleotide 3
zoom 200	magnifies

- To see this nucleotide, select "ball & stick" from the Display menu. Ignore the water molecules, in red, that also appear.
- Rotate the molecule until the gap between the deoxyribose rings of nucleotides 2 and 3 can be seen (P is in yellow).

restrict 2,3	restricts to nucleotides 2 and 3 in preparation for viewing phosphodiester bond formation

- Now view the formation of a phosphodiester bond between nucleotides 2 and 3 by choosing "ball & stick" from the Display menu. The bond will form in the gap between the two deoxyribose rings of nucleotides 2 and 3 as soon as "ball & stick" is selected, so make sure the pull-down menu is not covering the gap.

select 4	

- To view, choose "ball & stick" from the Display menu. Ignore the water molecules, in red, that also appear.
- Rotate the molecule until the gap between the deoxyribose rings of nucleotides 3 and 4 can be seen (P is in yellow).

restrict 2,3,4	restricts view to nucleotides 2, 3, and 4 in preparation for viewing phosphodiester bond formation

- Now view the phosphodiester bond formation between nucleotides 3 and 4 by using the Display menu to choose "ball & stick." The bond will form in the gap between the two deoxyribose residues of nucleotides 3 and 4.
    1. Between which two atoms does the phosphodiester bond form? ____

    _____

    2. To which two carbons of the deoxyribose rings is the phosphodiester bond attached? _____

    _____

zap	closes the file

Part IV: The Antiparallel Nature of DNA Strands

- Open the file "dna1.pdb" and view in "ball & stick" with "cpk" colors and the heteroatoms turned off.

restrict 2,3,4,21,22,23	restrict the view to these nucleotides
zoom 150	magnifies the view
select purine, pyrimidine	this will help distinguish the bases from the backbone
color yellow	
select backbone	

- Select "cpk" from the Color menu (the bases will remain yellow and the sugar-phosphate backbones will have cpk colors).
- Examine the terminal nucleotides on each chain, C21 and C23 on one chain and G4 and G2 on the other chain. Look at the terminal atoms of each, the ones that are not bonded to anything. One will be a phosphate group, and the other will be a hydroxyl group. (Note that the H of the hydroxyl group is not shown, only the O.)
- Identify the terminal group at each end of each chain and the carbons on the deoxyribose ring to which they are attached. The phosphate and hydroxyl groups (PO_4, OH) are named according to the number of the carbon ($1'$, $2'$, $3'$, $4'$, or $5'$) to which they are attached (e.g., $2'$-OH, $3'$-OH, or $5'$-PO_4).
    1. What is the terminal group on the C21 end? _____

    _____

    2. What is the terminal group on the C23 end? _____

    _____

    3. What is the terminal group on the G4 end? _____

    _____

    4. What is the terminal group on the G2 end? _____

    _____

- The differences in the terminal groups result from the orientation of the two strands. One strand is oriented in the opposite direction to the other strand. Rotate this short strand of DNA on the screen so that the purines and pyrimidines are oriented 90° to the plane of the screen (so only the edge of them is seen). Note that the oxygen (red) of the deoxyribose ring is oriented "up" on one strand and "down" on the other strand.

zap	closes the file

Part V: Interaction Between DNA and the Cro Repressor Protein

- Open the file "cro.pdb." This is a short stretch of DNA with two molecules of Cro protein attached to it.
- Select "ribbons" under the Display menu. Rotate the molecule until you see the double helix of DNA on one side and the two proteins on the other side. The proteins are small globular (round) structures with α-helices.
- Identify the major groove and the minor groove of the DNA. These appear as hollows or curves where there are no molecules. If each hollow around the three-dimensional structure is traced, you will find that both the large hollow (the major groove) and the smaller hollow (the minor groove) trace a helical path.
- Select "sticks" under the Display menu.

select a	selects A residues
color cyan	colors them cyan
select g	selects G residues
color green	colors them green
select t	selects T residues
color red	colors them red
select c	selects C residues
color magenta	colors them magenta

The nucleotides are now colored so that they can be easily identified.

select protein	selects the Cro proteins
color blue	colors them blue

- Select "sticks" under the Display menu again.

select all	
hbonds on	selects all of the hydrogen bonds
color hbonds yellow	colors all of them yellow
zoom 150	magnifies the view

TABLE 6.1
Properties of Amino Acid Side Chains

Properties of Side Chains (R Groups)	Amino Acids
Nonpolar	gly, ala, val, leu, ile, met, phe, trp, pro
Polar	ser, thr, cys, tyr, asn, gln
Electrically charged	asp, glu, lys, arg, his

- Rotate the molecule. Look for close interactions between the amino acids in the protein, colored blue, with the nucleotides of DNA, colored red, cyan, green, or magenta.
  1. Identify an amino acid that is in close proximity to and may interact with a nucleotide. Click on each with the mouse and identify the name and number of each molecule of the pair (one amino acid and one nucleotide). _____

     _____

  2. Move to another region of protein/DNA interaction and find another possible pair of amino acid and nucleotide that may interact and give the name and number of each. _____

     _____

  3. What generalization (i.e., polar, nonpolar, electrically charged) can be made about the amino acids that are closest to the DNA? (See Table 6.1.)

     _____

     _____

Cro interacts with DNA as a dimer, i.e., two identical polypeptides that bind together. The two Cro polypeptides that are bound to the DNA are examined next.

hbonds off	removes the hbonds
restrict protein	removes the nucleic acid
select helix	selects α helices of the protein (*not* DNA)
color yellow	colors them yellow

- Display as "ribbons."
- Rotate the two molecules of Cro. Try to see the orientation of each protein molecule relative to the other. If this cannot easily be seen, imagine that your closed fists represent the two Cro proteins. This represents the orientation of the two Cro proteins relative to each other. They are identical but rotated 180° from each other.

To see this more easily, some amino acids are next identified and colored in each Cro protein.

select lys14	selects amino acid 14, which is a lysine
color red	colors lys14 red in each Cro protein
select lys40	
color magenta	colors lys40 in each protein

- Note the relative positions of these amino acids in each molecule of Cro. Try selecting and coloring asp55, cys54, thr39, glu35, and val26, or click on amino acids with the mouse and look on the Command Line to get their name and number. (Available colors: red, green, blue, cyan, yellow, orange, white, purple, magenta.)

Next, the interaction of some of these amino acids with the DNA is examined.

select nucleic	selects the DNA
color white	colors the DNA white

- Display as "spacefill." The amino acids and their interaction with the major and minor groove should now be visible.
    4. To which groove does Cro bind? _____

    _____

select all	

- Display in "spacefill" to see more clearly how the Cro protein fits into the grooves of the DNA. Heteroatoms can be switched off.

restrict protein	
color blue	

- Display as "sticks."

select helix	
color yellow	

- Display as "ribbon."

select 28-36	selects amino acids 28-36 in Cro proteins
color green	
select nucleic	select the DNA
color white	

- Display as "spacefill."
- Rotate the molecule to see how one of the helices of each molecule of Cro fits in the major groove. To see the helix in the groove, you should be looking down the spiral center of the green helix of Cro, not from the side of the helix.

select all	

- Display as "spacefill." Heteroatoms can be switched off.

LAB 7

Isolation of Chromosomal DNA from *Bacillus licheniformis*

GOAL

The goal of this lab is to isolate pure, high molecular weight chromosomal DNA from *Bacillus licheniformis*.

OBJECTIVES

After completing Lab 7, you will be able to

1. isolate DNA from *Bacillus*
2. explain the chemical basis for DNA extraction
3. assess the quality and quantity of DNA by gel electrophoresis

BACKGROUND

Genes are functional units of DNA located in chromosomes. The α-amylase gene (*amyE*) is part of the chromosome in many strains of *Bacillus*. This bacterial chromosome is circular with a size of about 4000 kilobases (kb). The *amyE* gene is about 1.6 kb. In order to isolate the *amyE* gene of *Bacillus licheniformis*, first you must isolate chromosomal DNA from this strain.

ISOLATION OF DNA

The isolation of DNA from bacterial cells is relatively straightforward. The cells are broken open to release the DNA, proteins, and other cellular components. The DNA can be purified from the other components by extraction with phenol and chloroform. These are organic solvents that dissolve and denature proteins, dissolve lipids, and remove some polysaccharides. The DNA can be precipitated out of the aqueous phase by ethanol in the presence of a moderately high concentration of monovalent cations (e.g., Na^+). The precipitated DNA is dissolved in a buffer solution containing EDTA, which chelates, or binds, divalent cations. Since enzymes that degrade DNA (i.e., nucleases) require divalent cations, such as Mg^{2+}, for activity, the presence of EDTA minimizes DNA degradation.

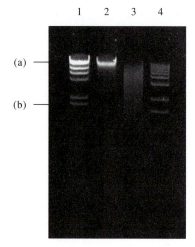

FIGURE 7.1

Resolution of DNA by agarose gel electrophoresis. Lane 1: λ DNA/*Hin*d III molecular weight markers—(*a*) 23.1 kb, (*b*) 2.0 kb; lane 2: high molecular weight *B. licheniformis* chromosomal DNA; lane 3: degraded *B. licheniformis* chromosomal DNA; lane 4: 1 kb DNA Ladder molecular weight markers (most bands differ in size by 1 kb).

ELECTROPHORESIS OF DNA

Fragments of DNA can be separated by electrophoresis through agarose gels. Agarose is a derivative of agar, a polysaccharide isolated from some seaweed. When a hot solution of agarose cools, the polysaccharide strands fold and twist around each other, forming a mesh or gel. The sample of DNA is put into a well at one end of the gel. When a current is applied, the negatively charged DNA molecules move through the gel toward the positive electrode. The semisolid gel has pores that hinder movement of the DNA fragments. Smaller size fragments of DNA move through the gel more quickly than do larger size fragments. The result is that DNA fragments separate on the basis of size (Fig. 7.1).

One common way to visualize DNA in agarose gels is by staining with ethidium bromide, a dye that fluoresces orange when exposed to UV light. Ethidium bromide is a flat planar molecule that intercalates, or slides between, the stacked base pairs of DNA. Since the fluorescence emitted by ethidium bromide bound to DNA is greater than that of unbound dye, small amounts of DNA can be detected in the presence of free ethidium bromide in the gel. If DNA molecular weight standards, which are linear fragments of DNA of known sizes, are run on the same gel, then the sizes of other fragments of DNA can be determined (see Fig. 7.1).

QUANTITATION OF DNA

Since the amount of fluorescence emitted by ethidium bromide bound to DNA is proportional to the total mass of DNA, the amount of DNA in a band on a gel can be estimated by comparing its intensity of staining with that of a DNA mass standard. The amount of DNA in the sample could also be estimated by taking absorbance readings at 260 nm and 280 nm. Nucleic acids, including DNA, absorb maximally at 260 nm. Proteins absorb maximally at 280 nm. A reading of 1 at 260 nm corresponds to a concentration of about 50 μg/ml of double-stranded DNA. The ratio between readings at 260 nm and 280 nm provides an estimate of the purity of the sample. Pure preparations of nucleic acids have A_{260}/A_{280} ratios of 1.8 to 2.0. If the sample contains protein or other contaminants (e.g., phenol), however, then the ratio is much lower, and it is not possible to accurately quantitate the amount of DNA with this method.

LABORATORY OVERVIEW

This lab begins the section of the course in which you will isolate the *amyE* gene of *B. licheniformis* and clone it into *E. coli*. In this lab, you will isolate chromosomal DNA from *B. licheniformis* cells. In Lab 8, you will use this DNA as a template to amplify and label with biotin a portion of the *amyE* gene, using the polymerase chain reaction (PCR). This biotin-labeled fragment will be used in Lab 9 as a probe to identify restriction fragments of *B. licheniformis* chromosomal DNA that contain the *amyE* gene. One of these restriction fragments will be used to clone the *amyE* gene in Lab 10.

For your cloning project to be a success, it is important that you isolate high-quality chromosomal DNA; i.e., high molecular weight DNA that is free of protein and other contaminants that could interfere with digestion by restriction enzymes. The critical step in obtaining high-quality chromosomal DNA from *B. licheniformis* is to prevent degradation by nucleases during cell lysis. Thus, you will pulverize the cells with glass beads in the presence of phenol and chloroform so that proteins (e.g., nucleases) will be denatured as soon as the cells are broken open. After centrifugation, the denatured proteins appear as a white layer between the aqueous and organic phases. The aqueous phase containing the DNA is removed and purified further by additional extractions with phenol and chloroform. These extractions will be performed in the presence of a proprietary compound called Phase Lock Gel™, which, upon centrifugation, forms a physical barrier between the two phases. The organic phase and the interface material are trapped below the Phase Lock Gel (PLG), while the DNA-containing aqueous phase remains above. Use of this compound minimizes your exposure to phenol and chloroform and maximizes the purity of the DNA.

Even though you will perform your organic extractions in the presence of PLG, your final sample of chromosomal DNA will most likely contain enough phenol to interfere with absorbance readings. Thus, you will estimate the amount of DNA in your preparation by running a small amount on an agarose gel and comparing its intensity of ethidium bromide staining with that of a DNA mass standard. Electrophoresis will also allow you to estimate the size and assess the quality of your chromosomal DNA. High molecular weight DNA migrates very slowly, and your purified chromosomal DNA should appear as a distinct band near the top of the gel.

In the exercises at the end of this lab, you will examine the relationship between the sequence of nucleotides in DNA and the sequence of amino acids that may be encoded by that DNA. You will identify the correct amino acid translation of the *amyE* gene from *Bacillus licheniformis.*

TIMELINE

DNA isolation takes about 2–2.5 hours. Agarose gel electrophoresis takes 2–3 hours (depends upon whether the gels are cast prior to the lab session).

SAFETY GUIDELINES

Phenol is a powerful corrosive and can quickly cause serious chemical burns. It damages cells by precipitating and denaturing proteins. Phenol is a respiratory irritant and readily penetrates the skin; it can damage the nervous system, liver, and kidneys. Some people get severe headaches when inhaling only small amounts of phenol vapors. Wear gloves, eye protection, a lab coat, and always work in a fume hood when using phenol.

Chloroform is carcinogenic and can cause organ damage. It can be inhaled or absorbed through the skin. Its effects are cumulative. Chloroform, when mixed with phenol, enhances the ability of phenol to penetrate the skin. Wear gloves, eye protection, a lab coat, and always work in a fume hood when using chloroform.

Boiling agarose can cause burns. Wear protective gloves when handling hot agarose solutions.

The electric current in a gel electrophoresis device is extremely dangerous. Never remove a lid or touch the buffer once the power is turned on. Make sure that the counter where the gel is being run is dry.

Ethidium bromide is a strong mutagen, by virtue of its insertion between the bases of DNA. Gloves should always be worn when handling gels or buffers containing this chemical.

UV light, used to illuminate the DNA stained with ethidium bromide, is dangerous. Protect your eyes and face by wearing a UV-blocking face shield.

PROCEDURE

LYSE THE CELLS

☐   1.   Transfer about 25 ml of an overnight culture of *B. licheniformis* to an appropriately sized centrifuge tube. Centrifuge at about 3500 × g (e.g., 5500 rpm in a Sorvall SS34 rotor) for 5 minutes to pellet the cells.

☐   2.   Pour off the culture medium into a container for liquid biological waste. Add 2 ml of TE (10 mM Tris-HCl, pH 8, 1 mM EDTA) to the pellet of cells and vortex vigorously. Check whether the cells have been resuspended uniformly by tipping the tube upside down to disperse the mixture along the walls of the tube. If any clumps of cells are present, continue to vortex until you have a homogeneous suspension. Add ~20 ml of TE, and vortex to wash the cells. The purpose of this step is to wash away the culture medium, and it is much easier to resuspend the cells uniformly by first mixing them with a small volume of buffer. Centrifuge at $3500 \times g$ for 5 minutes to pellet the cells again.

☐   3.   Pour off and discard the wash solution. Add 2 ml of TE, and vortex vigorously to resuspend the cells, as described in step 2. Place the tube of cells on ice until step 5.

☐   4.   Transfer 1 ml of glass beads to a 15-ml, round-bottom centrifuge tube. For this, use a 5-ml serological pipette and a pipette pump to pull approximately 4–5 ml of the glass bead suspension into the pipette. Allow the glass beads to settle to the bottom of the pipette (liquid will remain on top of the column of beads). With the pipette tip still in the bead container, dispense the beads slowly and carefully until the level of the beads is at a major marking (e.g., 2 or 3 ml). Position the pipette in the bottom of the 15-ml tube and dispense 1 ml of the beads. Do not add any liquid.

☐   5.   Use a pipette to transfer all of the resuspended cells (from step 3) to the 15-ml centrifuge tube containing the glass beads.

> ✳ CAUTION: Perform all transfers and operations that involve phenol or chloroform in a fume hood. You must wear gloves, lab coat, and eye protection.

☐   6.   Add 1 ml of phenol:chloroform:isoamyl alcohol (25:24:1) to the tube of cells and glass beads. (Isoamyl alcohol is an antifoaming agent.) Close the tube securely to prevent leakage during the subsequent steps.

☐   7.   Vortex for 1 minute, then place the tube on ice for 1 minute. Repeat this sequence three more times to give a total vortexing time of 4 minutes. Vortexing in the presence of glass beads and organic compounds causes the cells to lyse and release DNA, proteins (which are denatured by the phenol and chloroform), and other cellular components.

> ✳ CAUTION: If the contents of the tube leak and get on your gloves, change your gloves, as phenol will go through the gloves to your skin. If your skin feels tingly, then phenol has come in contact with it. Remove and discard your gloves and immediately wash your hands with soap and water. If the tube has leaked, make sure the rubber top on the vortex mixer is cleaned thoroughly.

☐   8.   Centrifuge the lysed cell suspension at $6000 \times g$ for 5 minutes to separate the phases. *Note:* You may have to put your tube into a rubber adapter so that it fits in the centrifuge rotor.

9. Remove the tube containing the cell lysate very carefully from the centrifuge rotor and place it in a rack. There will be three distinct layers in the tube: the bottom layer contains the glass beads and organic compounds, the middle opaque layer contains denatured proteins and other cellular debris, and the upper clear (aqueous) layer contains the DNA. Do not jostle the tube or you will disturb the layers.

EXTRACT THE DNA

1. Prepare a 15-ml Phase Lock Gel (PLG) tube by centrifuging it at $1500 \times g$ for 2 minutes.

2. Carefully transfer 2 ml of the upper clear aqueous layer from the cell lysate tube (from step 9 in the preceding section) to the PLG tube. Do not transfer any of the cloudy material near the middle interface layer. Discard the tube with phenol, beads, and cell debris into a container marked for waste phenol.

    The DNA in the aqueous phase will be purified by two more extractions with phenol and chloroform and a final extraction with chloroform.

3. ***TAKE APPROPRIATE PRECAUTIONS WHEN HANDLING PHENOL AND CHLOROFORM.*** Add 2 ml of phenol:chloroform: isoamyl alcohol (25:24:1) to the PLG tube. Cap and gently invert the tube to mix the aqueous and organic phases until a cloudy emulsion forms. Do not vortex the PLG tube. The PLG will not become a part of the mixture; it will remain at the bottom of the tube.

4. Centrifuge the PLG tube at $10,000 \times g$ for 5 minutes to separate the phases. The PLG will migrate up the tube and form a physical barrier between the upper aqueous phase, containing the DNA, and the lower organic phase, containing denatured protein and other contaminants.

5. Add another 2 ml of phenol:chloroform:isoamyl alcohol (25:24:1) to the PLG tube. Cap and gently invert the tube to mix the aqueous phase and the newly added organic phase until a cloudy emulsion forms above the PLG layer, which will not move. Do not vortex the tube.

6. Centrifuge the PLG tube again at $10,000 \times g$ for 5 minutes to separate the phases. The PLG will migrate up the tube to form a new physical barrier between the aqueous and organic phases.

7. Add 2 ml of chloroform:isoamyl alcohol (24:1) to the PLG tube. Pipette carefully—chloroform has a tendency to drip out of pipette tips. This extraction with chloroform removes traces of phenol from the DNA sample. Cap and thoroughly mix to form a homogenous suspension above the plug of PLG. Do not vortex the PLG tube.

8. Centrifuge at $10,000 \times g$ for 5 minutes to separate the phases; the PLG will migrate up the tube and form a new barrier between the upper aqueous phase and the lower organic phase.

9. Transfer 1.5 ml of the upper aqueous phase containing the DNA from the PLG tube to a new 15-ml tube. Discard the PLG tube into a container marked for waste phenol.

PRECIPITATE THE DNA

1. Add 150 μl (0.1 volume) of 3 M sodium acetate (pH 5.2) to the 15-ml tube with the DNA solution. Cap and invert the tube to mix. *Note:* If you forget to add the sodium acetate, the DNA will not precipitate when you add the ethanol.

☐   2.   Add 3.3 ml (2 volumes) of ice-cold 95% ethanol. Cap and slowly invert the tube to mix. You should see the DNA precipitate out of solution as long white strands (i.e., it looks like cotton candy).

☐   3.   You will collect the strands of DNA by "hooking" them onto a glass Pasteur pipette with a U-shaped tip. First, prepare the Pasteur pipette by inserting its tip, horizontally, into the flame of a Bunsen burner. As soon as the glass melts to seal the tip, the tip turns downward and forms a little hook. Remove from the flame and let cool. *Note:* The hook cannot be too large; it must fit into the bottom of a 1.5-ml microcentrifuge tube.

Now, collect the strands of precipitated DNA with the hook of your pipette. It may help to swirl the pipette to collect the DNA. Don't worry about any small pieces of DNA that may be floating in the ethanol solution; just gather the long strands.

☐   4.   Place the pipette with the hooked DNA into a 1.5-ml microcentrifuge tube containing 1 ml of 70% ethanol and let it sit for about 2 minutes. The DNA will remain precipitated and stuck to the hook of the pipette, but much of the sodium acetate will be removed from the DNA.

☐   5.   Carefully remove the pipette with the precipitated DNA from the tube of 70% ethanol and invert the pipette tip upright for a minute or two until the excess ethanol has drained away and evaporated. Do not let the DNA dry completely.

☐   6.   Place the pipette with the precipitated DNA into a 1.5-ml microcentrifuge tube containing 120 μl of TE. Let the pipette sit in the TE until the DNA slides off the hook. Cap the tube and label as "*Bacillus* DNA," with your initials or group number and the date.

☐   7.   Let the DNA dissolve completely in the TE by incubating it overnight at 4°C. After the DNA has dissolved, it can be stored at −20°C.

CHECK THE DNA BY GEL ELECTROPHORESIS

☐   1.   Cast a 0.7% agarose gel, as described in the section on Agarose Gel Electrophoresis in Appendix II. Once it has solidified, cover the gel with 1 × Tris Acetate EDTA (TAE) Buffer (40 mM Tris-acetate, 1 mM EDTA).

☐   2.   To assess the quality and quantity of your chromosomal DNA, you must dilute some to run on the gel. Add 45 μl of TE to each of two 1.5-ml microcentrifuge tubes. Label the first as "10^{-1} dilution" and the second as "10^{-2} dilution." To the first tube, add 5 μl of your chromosomal DNA (Fig. 7.2). The solution of chromosomal DNA will be viscous and difficult to pipette. Hold the tubes at eye level when pipetting to or from them. Watch the pipette tip fill, keeping the tip immersed until the liquid stops rising in the tip. Flush out the tip by pipetting back and forth when transferring the DNA to the tube with TE. Vortex the "10^{-1} dilution" tube. Add 5 μl of this dilution to the "10^{-2} dilution" tube, as previously described. Vortex the "10^{-2} dilution" tube.

☐   3.   Transfer 5 μl of each dilution of DNA to new, labeled 1.5-ml tubes. Add 5 μl of 2 × DNA Gel Loading Buffer (0.08% Bromophenol Blue, 16.7 mM EDTA, 13.3% sucrose) to each tube, and mix by pipetting up and down. (The sucrose increases the density of the sample so it can be loaded in the wells of the gel, and the blue dye permits tracking of the sample during electrophoresis.) Centrifuge the tubes briefly to collect the entire sample in the bottom of each tube.

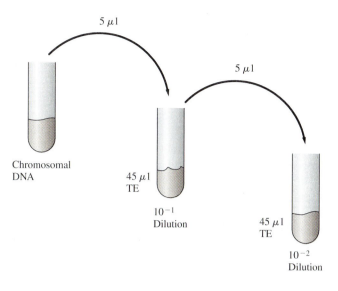

FIGURE 7.2
Serial dilution of chromosomal DNA.

4. Before loading the DNA samples on the gel, each student should practice loading one lane of the gel with 10 μl of DNA Gel Loading Buffer. The loading buffer can be flushed out of the wells with a Pasteur pipette.

5. Load the entire 10 μl volume of each DNA sample into separate wells of the gel. Load DNA molecular weight markers on the gel. (Your instructor will tell you how much to load and tell you the sizes of the standards you are using.) Keep a record of which sample was loaded in which lane.

6. Place the lid on the gel chamber, and connect the positive electrode to the bottom end of the gel and the negative electrode to the top (well-end) of the gel. Run at 80–90 volts for 1.5–2 hours or until the blue dye has migrated about three-quarters of the way to the bottom of the gel.

7. Read about staining gels and working with ethidium bromide in the Agarose Gel Electrophoresis section of Appendix II. **WEAR GLOVES WHEN HANDLING SOLUTIONS OF ETHIDIUM BROMIDE, AND WEAR EYE PROTECTION WHEN USING UV LIGHT.** Stain the gel with ethidium bromide. Photograph the gel using a Polaroid camera with an orange filter. Label and date the photograph and place it in your lab notebook.

DATA ANALYSIS

- Label each lane and denote the size of each DNA molecular weight marker on the photograph (or photocopy) of your gel.
- Since you broke open the cells by pulverizing them with glass beads, the chromosomal DNA has been broken into fragments of about 25–50 kb. These high molecular weight fragments of chromosomal DNA migrate very slowly and will appear as a distinct band near the top of the gel (see lane 2 of Fig. 7.1). If your chromosomal DNA is degraded, it will appear as a smear along the length of the gel (see lane 3 of Fig. 7.1). This smear is due to the thousands of different sized fragments that merge along the gel to create a continuous smear rather than discrete bands.
- Use the lane with the 10^{-1} dilution of DNA to assess the quality of your chromosomal DNA. Determine whether the majority of it is high molecular weight or whether most of it is degraded. This lane should contain a

thick, bright band of high molecular weight DNA near the top of the gel. Even if there are some smaller-sized fragments below your thick, bright band near the top of the gel, your DNA is still mainly high molecular weight. If, however, there is no distinct high molecular weight band but only a smear along the length of the gel, then your DNA is degraded. If your DNA appears degraded, do not use it for Labs 8, 9, 10. Borrow some DNA from a group whose DNA does not appear to be degraded. There may also be a bright diffuse band ($<$0.5 kb) near the bottom of the lane. This is rRNA and tRNA that co-purified with the DNA. It is generally disregarded and of little consequence in most samples of DNA (but it can interfere with the detection of small fragments of DNA). It can be removed by treatment with RNase, a nuclease that recognizes and degrades RNA but not DNA.

- Use the lane with the 10^{-2} dilution of DNA to estimate the concentration of your preparation of chromosomal DNA. First, determine the amount of DNA in that band by comparing its intensity of staining with that of one of the larger DNA molecular weight standards. If, for example, your band of 10^{-2} diluted DNA is about twice as bright as the band of standard DNA and the standard band contains 10 ng of DNA, then your band of 10^{-2} diluted DNA contains 2×10 ng $= 20$ ng. Second, determine the concentration of DNA in your diluted sample. If 5 μl of 10^{-2} diluted DNA was loaded in that lane (ignore the 5 μl of loading dye), then the concentration of the diluted DNA would be 20 ng $\div$ 5 μl $= 4$ ng/μl. Third, determine the concentration of DNA in your undiluted sample of chromosomal DNA. Since you diluted the DNA a hundredfold, then its original concentration would be 4 ng/μl $\times$ 100 $= 400$ ng/μl. Write the DNA concentration on your tube of chromosomal DNA and record this value in your notebook.

- Calculate your total yield of DNA (in μg). To determine this, multiply the concentration of your undiluted sample of chromosomal DNA by the volume of your sample (i.e., the amount of TE used to dissolve your precipitated DNA). For example,

$$400 \text{ ng/}\mu\text{l} \times 120 \text{ }\mu\text{l} = 48{,}000 \text{ ng} \times 1 \text{ }\mu\text{g/1000 ng} = 48 \text{ }\mu\text{g of DNA}$$

QUESTIONS

1. Why did you break open the *B. licheniformis* cells by vortexing them with glass beads in the presence of phenol and chloroform?
2. Explain the chemical basis for extraction of DNA with phenol and chloroform.
3. Why must all manipulations with phenol and chloroform be performed in a fume hood?
4. What happens to DNA in the presence of 0.3 M sodium acetate (pH 5.2) and 60–70% ethanol?
5. Explain how ethidium bromide interacts with DNA.
6. What would happen to your DNA samples if the positive and negative leads on the chamber were reversed at the start of electrophoresis (i.e., the negative electrode was connected to the bottom of the gel)?

EXERCISES

The sequence of nucleotides in a DNA molecule represents the genetic information. This nucleotide sequence can be used to deduce the amino acid sequence of a protein that may be encoded by the DNA. The relationship between the sequence of nucleotides and the sequence of the corresponding protein is called the *genetic code.* The genetic code is read in groups of three nucleotides, with each trinucleotide sequence (codon) corresponding to one of the 20 amino acids. The genetic code also contains three codons that stop translation (termination codons).

The nucleotide sequence is always read in the 5′ to 3′ direction and encodes a polypeptide from the amino terminus to the carboxyl terminus. There are three possible ways of reading a nucleotide sequence as a series of triplets, depending upon the starting point. These are called *reading frames.* A reading frame consisting solely of triplets that represent amino acids is called an *open reading frame* (ORF). A complete ORF should have a start codon (almost always an ATG that encodes methionine), triplets that encode an uninterrupted sequence of amino acids, and a termination codon (TAA, TAG, or TGA). Since either strand of DNA can code for protein, a region of double-stranded DNA has six different reading frames (i.e., each strand has three reading frames).

Generally, computer software programs are used to translate nucleotide sequences into amino acid sequences and identify potential ORFs. Read the section on Bioinformatics in Appendix I and perform the Sequence Analysis Using Computer Software exercise in Appendix I. In the following two exercises, you will translate a short sequence of DNA into amino acids and examine computer-generated amino acid translations of the *amyE* gene of *B. licheniformis.*

EXERCISE 7.1
Translate DNA into Protein

A short sequence of double-stranded DNA is presented in this exercise. Use the genetic code in Table 7.1 to translate this short sequence of DNA (5′ to 3′) into a sequence of amino acids in all six reading frames. Find the amino acid (in Table 7.1) that corresponds to each codon in each reading frame and record it below. If you find a termination (stop) codon, indicate this with an asterisk. The first amino acid in each reading frame is presented for you.

Reading Frame 1   Ser

Reading Frame 2   Gln

Reading Frame 3   Arg

```
          1         5            10            15        20            25
     5′-T C A G A T G A C T T A C A G C C A T A A A C G T-3′
     3′-A G T C T A C T G A A T G T C G G T A T T T G C A-5′
```

Reading Frame 4   Thr

Reading Frame 5   Arg

Reading Frame 6   Val

1. Which reading frame(s) contain termination codons? _____
   _____

2. Which reading frame, if any, contains an amino acid that might be the start of a protein? _____
   _____

TABLE 7.1
The Genetic Code (The one-letter symbol for each amino acid is presented in parentheses.)

First Nucleotide Position	Second Nucleotide Position				Third Nucleotide Position
	T	**C**	**A**	**G**	
T	Phe (F)	Ser (S)	Tyr (Y)	Cys (C)	T
	Phe	Ser	Tyr	Cys	C
	Leu (L)	Ser	Stop	Stop	A
	Leu	Ser	Stop	Trp (W)	G
C	Leu (L)	Pro (P)	His (H)	Arg (R)	T
	Leu	Pro	His	Arg	C
	Leu	Pro	Gln (Q)	Arg	A
	Leu	Pro	Gln	Arg	G
A	Ile (I)	Thr (T)	Asn (N)	Ser (S)	T
	Ile	Thr	Asn	Ser	C
	Ile	Thr	Lys (K)	Arg (R)	A
	Met (M)	Thr	Lys	Arg	G
G	Val (V)	Ala (A)	Asp (D)	Gly (G)	T
	Val	Ala	Asp	Gly	C
	Val	Ala	Glu (E)	Gly	A
	Val	Ala	Glu	Gly	G

EXERCISE 7.2
Find an Open Reading Frame

On the following pages, Figure 7.3 presents the DNA sequence and three amino acid translations of a region of DNA that contains the *amyE* gene of *Bacillus licheniformis*. This sequence (accession # X03236) was obtained from the GenBank DNA sequence database at the National Center for Biotechnology Information at the National Institutes of Health. The three reading frames of the upper strand are presented, using the one-letter amino acid symbols (see Table 7.1). A dot indicates a stop codon. Examine the three reading frames of translation of the *amyE* gene of *Bacillus licheniformis* and answer the following questions.

1. Which reading frame (1, 2, or 3) has a complete ORF that most likely encodes α-amylase? _____

2. List the positions of two potential start codons in the ORF you just identified. (Give the location of the first nucleotide of the codon.) Which amino acid does this codon encode? _____

3. Given that ribosomes bind to A/G-rich regions about seven nucleotides upstream from the start codon, which potential start codon is most likely the correct one? _____

4. Give the sequence and position of the termination codon for the ORF. _____

5. How many amino acids are there in this ORF? _____

6. Using the mean molecular weight of an amino acid, calculate the estimated size of the α-amylase protein of *B. licheniformis*. _____

```
ATTGGTAACTGTATCTCAGCTTGAAGAAGTGAAGAAGCAGAGAGGCTATTGAATAAATGAGTAGAAAGCGCCATATCGGCGCTTTTCTTTTGGAAGAAAA
+----+----+----+----+----+----+----+----+----+----+----+----+----+----+----+----+----+----+----+----+  100
TAACCATTGACATAGAGTCGAACTTCTTCACTTCTTCGTCTCTCCGATAACTTATTTACTCATCTTTCGCGGTATAGCCGCGAAAAGAAAACCTTCTTTT

   I  G  N  C  I  S  A  .  R  S  E  E  A  E  R  L  L  N  K  .  V  E  S  A  I  S  A  L  F  F  W  K  K
    L  V  T  V  S  Q  L  E  E  V  K  K  Q  R  G  Y  .  I  N  E  .  K  A  P  Y  R  R  F  S  F  G  R  K
   Y  W  .  L  Y  L  S  L  K  K  .  R  S  R  E  A  I  E  .  M  S  R  K  R  H  I  G  A  F  L  L  E  E  N

TATAGGGAAAATGGTATTTGTTAAAAATTCGGAATATTTATACAATATCATATGTTTCACATTGAAAGGGGAGGAGAATCATGAAACAACAAAAACGGCT
+----+----+----+----+----+----+----+----+----+----+----+----+----+----+----+----+----+----+----+----+  200
ATATCCCTTTTACCATAAACAATTTTTAAGCCTTATAAATATGTTATAGTATACAAAGTGTAACTTTCCCCTCCTCTTAGTACTTTGTTGTTTTTGCCGA

   I  .  G  K  W  Y  L  L  K  I  R  N  I  Y  T  I  S  Y  V  S  H  .  K  G  R  R  I  M  K  Q  Q  K  R  L
    Y  R  E  N  G  I  C  .  K  F  G  I  F  I  Q  Y  H  M  F  H  I  E  R  G  G  E  S  .  N  N  K  N  G
     I  G  K  M  V  F  V  K  N  S  E  Y  L  Y  N  I  I  C  F  T  L  K  G  E  E  N  H  E  T  T  K  T  A

TTACGCCCGATTGCTGACGCTGTTATTTGCGCTCATCTTCTTGCTGCCTCATTCTGCAGCAGCGGCGGCAAATCTTAATGGGACGCTGATGCAGTATTTT
+----+----+----+----+----+----+----+----+----+----+----+----+----+----+----+----+----+----+----+----+  300
AATGCGGGCTAACGACTGCGACAATAAACGCGAGTAGAAGAACGACGGAGTAAGACGTCGTCGCCGCCGTTTAGAATTACCCTGCGACTACGTCATAAAA

    Y  A  R  L  L  T  L  L  F  A  L  I  F  L  L  P  H  S  A  A  A  A  N  L  N  G  T  L  M  Q  Y  F
   F  T  P  D  C  .  R  C  Y  L  R  S  S  S  C  C  L  I  L  Q  Q  R  R  Q  I  L  M  G  R  .  C  S  I  L
    L  R  P  I  A  D  A  V  I  C  A  H  L  L  A  A  S  F  C  S  S  G  G  K  S  .  W  D  A  D  A  V  F

GAATGGTACATGCCCAATGACGGCCAACATTGGAAGCGCTTGCAAAACGACTCGGCATATTTGGCTGAACACGGTATTACTGCCGTCTGGATTCCCCCGG
+----+----+----+----+----+----+----+----+----+----+----+----+----+----+----+----+----+----+----+----+  400
CTTACCATGTACGGGTTACTGCCGGTTGTAACCTTCGCGAACGTTTTGCTGAGCCGTATAAACCGACTTGTGCCATAATGACGGCAGACCTAAGGGGGCC

   E  W  Y  M  P  N  D  G  Q  H  W  K  R  L  Q  N  D  S  A  Y  L  A  E  H  G  I  T  A  V  W  I  P  P
    N  G  T  C  P  M  T  A  N  I  G  S  A  C  K  T  T  R  H  I  W  L  N  T  V  L  L  P  S  G  F  P  R
     .  M  V  H  A  Q  .  R  P  T  L  E  A  L  A  K  R  L  G  I  F  G  .  T  R  Y  Y  C  R  L  D  S  P  G

CATATAAGGGAACGAGCCAAGCGGATGTGGGCTACGGTGCTTACGACCTTTATGATTTAGGGGAGTTTCATCAAAAAGGGACGGTTCGGACAAAGTACGG
+----+----+----+----+----+----+----+----+----+----+----+----+----+----+----+----+----+----+----+----+  500
GTATATTCCCTTGCTCGGTTCGCCTACACCCGATGCCACGAATGCTGGAAATACTAAATCCCCTCAAAGTAGTTTTTCCCTGCCAAGCCTGTTTCATGCC

   A  Y  K  G  T  S  Q  A  D  V  G  Y  G  A  Y  D  L  Y  D  L  G  E  F  H  Q  K  G  T  V  R  T  K  Y  G
    H  I  R  E  R  A  K  R  M  W  A  T  V  L  T  T  F  M  I  .  G  S  F  I  K  K  G  R  F  G  Q  S  T
     I  .  G  N  E  P  S  G  C  G  L  R  C  L  R  P  L  .  F  R  G  V  S  S  K  R  D  G  S  D  K  V  R

CACAAAAGGAGAGCTGCAATCTGCGATCAAAAGTCTTCATTCCCGCGACATTAACGTTTACGGGGATGTGGTCATCAACCACAAAGGCGGCGCTGATGCG
+----+----+----+----+----+----+----+----+----+----+----+----+----+----+----+----+----+----+----+----+  600
GTGTTTTCCTCTCGACGTTAGACGCTAGTTTTCAGAAGTAAGGGCGCTGTAATTGCAAATGCCCCTACACCAGTAGTTGGTGTTTCCGCCGCGACTACGC

    T  K  G  E  L  Q  S  A  I  K  S  L  H  S  R  D  I  N  V  Y  G  D  V  V  I  N  H  K  G  G  A  D  A
   A  Q  K  E  S  C  N  L  R  S  K  V  F  I  P  A  T  L  T  F  T  G  M  W  S  S  T  T  K  A  A  L  M  R
    H  K  R  R  A  A  I  C  D  Q  K  S  S  F  P  R  H  .  R  L  R  G  C  G  H  Q  P  Q  R  R  R  .  C

ACCGAAGATGTAACCGCGGTTGAAGTCGATCCCGCTGACCGCAACCGCGTAATTTCAGGAGAACACCGAATTAAAGCCTGGACACATTTTCATTTTCCGG
+----+----+----+----+----+----+----+----+----+----+----+----+----+----+----+----+----+----+----+----+  700
TGGCTTCTACATTGGCGCCAACTTCAGCTAGGGCGACTGGCGTTGGCGCATTAAAGTCCTCTTGTGGCTTAATTTCGGACCTGTGTAAAAGTAAAAGGCC

   T  E  D  V  T  A  V  E  V  D  P  A  D  R  N  R  V  I  S  G  E  H  R  I  K  A  W  T  H  F  H  F  P
    P  K  M  .  P  R  L  K  S  I  P  L  T  A  T  A  .  F  Q  E  N  T  E  L  K  P  G  H  I  F  I  F  R
     D  R  R  C  N  R  G  .  S  R  S  R  .  P  Q  P  R  N  F  R  R  T  P  N  .  S  L  D  T  F  S  F  S  G

GGCGCGGCAGCACATACAGCGATTTTAAATGGCATTGGTACCATTTTGACGGAACCGATTGGGACGAGTCCCGAAAGCTGAACCGCATCTATAAGTTTCA
+----+----+----+----+----+----+----+----+----+----+----+----+----+----+----+----+----+----+----+----+  800
CCGCGCCGTCGTGTATGTCGCTAAAATTTACCGTAACCATGGTAAAACTGCCTTGGCTAACCCTGCTCAGGGCTTTCGACTTGGCGTAGATATTCAAAGT

   G  R  G  S  T  Y  S  D  F  K  W  H  W  Y  H  F  D  G  T  D  W  D  E  S  R  K  L  N  R  I  Y  K  F  Q
    G  A  A  A  H  T  A  I  L  N  G  I  G  T  I  L  T  E  P  I  G  T  S  P  E  S  .  T  A  S  I  S  F
     A  R  Q  H  I  Q  R  F  .  M  A  L  V  P  F  .  R  N  R  L  G  R  V  P  K  A  E  P  H  L  .  V  S

AGGAAAGGCTTGGGATTGGGAAGTTTCCAATGAAAACGGCAACTATGATTATTTGATGTATGCCGACATCGATTATGACCATCCTGATGTCGCAGCAGAA
+----+----+----+----+----+----+----+----+----+----+----+----+----+----+----+----+----+----+----+----+  900
TCCTTTCCGAACCCTAACCCTTCAAAGGTTACTTTTGCCGTTGATACTAATAAACTACATACGGCTGTAGCTAATACTGGTAGGACTACAGCGTCGTCTT

    G  K  A  W  D  W  E  V  S  N  E  N  G  N  Y  D  Y  L  M  Y  A  D  I  D  Y  D  H  P  D  V  A  A  E
   K  E  R  L  G  I  G  K  F  P  M  K  T  A  T  M  I  I  .  C  M  P  T  S  I  M  T  I  L  M  S  Q  Q  K
    R  K  G  L  G  L  G  S  F  Q  .  K  R  Q  L  .  L  F  D  V  C  R  H  R  L  .  P  S  .  C  R  S  R
```

FIGURE 7.3

Bacillus licheniformis amyE gene with upstream and downstream sequences.

```
ATTAAGAGATGGGGCACTTGGTATGCCAATGAACTGCAATTGGACGGTTTCCGTCTTGATGCTGTCAAACACATTAAATTTTCTTTTTTGCGGGATTGGG
+---+---+---+---+---+---+---+---+---+---+---+---+---+---+---+---+---+---+---+---+    1000
TAATTCTCTACCCCGTGAACCATACGGTTACTTGACGTTAACCTGCCAAAGGCAGAACTACGACAGTTTGTGTAATTTAAAAGAAAAAACGCCCTAACCC

  I  K  R  W  G  T  W  Y  A  N  E  L  Q  L  D  G  F  R  L  D  A  V  K  H  I  K  F  S  F  L  R  D  W
  L  R  D  G  A  L  G  M  P  M  N  C  N  W  T  V  S  V  L  M  L  S  N  T  L  N  F  L  F  C  G  I  G
  N  .  E  M  G  H  L  V  C  Q  .  T  A  I  G  R  F  P  S  .  C  C  Q  T  H  .  I  F  F  F  A  G  L  G

TTAATCATGTCAGGGAAAAAACGGGGAAGGAAATGTTTACGGTAGCTGAATATTGGCAGAATGACTTGGGCGCGCTGGAAAACTATTTGAACAAAACAAA
+---+---+---+---+---+---+---+---+---+---+---+---+---+---+---+---+---+---+---+---+    1100
AATTAGTACAGTCCCTTTTTTGCCCCTTCCTTTACAAATGCCATCGACTTATAACCGTCTTACTGAACCCGCGCGACCTTTTGATAAACTTGTTTTGTTT

  V  N  H  V  R  E  K  T  G  K  E  M  F  T  V  A  E  Y  W  Q  N  D  L  G  A  L  E  N  Y  L  N  K  T  N
  L  I  M  S  G  K  K  R  G  R  K  C  L  R  .  L  N  I  G  R  M  T  W  A  R  W  K  T  I  .  T  K  Q
  .  S  C  Q  G  K  N  G  E  G  N  V  Y  G  S  .  I  L  A  E  .  L  G  R  A  G  K  L  F  E  Q  N  K

TTTTAATCATTCAGTGTTTGACGTGCCGCTTCATTATCAGTTCCATGCTGCATCGACACAGGGAGGCGGCTATGATATGAGGAAATTGCTGAACAGTACG
+---+---+---+---+---+---+---+---+---+---+---+---+---+---+---+---+---+---+---+---+    1200
AAAATTAGTAAGTCACAAACTGCACGGCGAAGTAATAGTCAAGGTACGACGTAGCTGTGTCCCTCCGCCGATACTATACTCCTTTAACGACTTGTCATGC

  F  N  H  S  V  F  D  V  P  L  H  Y  Q  F  H  A  A  S  T  Q  G  G  G  Y  D  M  R  K  L  L  N  S  T
  I  L  I  I  Q  C  L  T  C  R  F  I  I  S  S  M  L  H  R  H  R  E  A  A  M  I  .  G  N  C  .  T  V  R
  F  .  S  F  S  V  .  R  A  A  S  L  S  V  P  C  C  I  D  T  G  R  R  L  .  Y  E  E  I  A  E  Q  Y

GTCGTTTCCAAGCATCCGTTGAAAGCGGTTACATTTGTCGATAACCATGATACACAGCCGGGGCAATCGCTTGAGTCGACTGTCCAAACATGGTTTAAGC
+---+---+---+---+---+---+---+---+---+---+---+---+---+---+---+---+---+---+---+---+    1300
CAGCAAAGGTTCGTAGGCAACTTTCGCCAATGTAAACAGCTATTGGTACTATGTGTCGGCCCCGTTAGCGAACTCAGCTGACAGGTTTGTACCAAATTCG

  V  V  S  K  H  P  L  K  A  V  T  F  V  D  N  H  D  T  Q  P  G  Q  S  L  E  S  T  V  Q  T  W  F  K
  S  F  P  S  I  R  .  K  R  L  H  L  S  I  T  M  I  H  S  R  G  N  R  L  S  R  L  S  K  H  G  L  S
  G  R  F  Q  A  S  V  E  S  G  Y  I  C  R  .  P  .  Y  T  A  G  A  I  A  .  V  D  C  P  N  M  V  .  A

CGCTTGCTTACGCTTTTATTCTCACAAGGGAATCTGGATACCCTCAGGTTTTCTACGGGGATATGTACGGGACGAAAGGAGACTCCCAGCGCGAAATTCC
+---+---+---+---+---+---+---+---+---+---+---+---+---+---+---+---+---+---+---+---+    1400
GCGAACGAATGCGAAAATAAGAGTGTTCCCTTAGACCTATGGGAGTCCAAAAGATGCCCCTATACATGCCCTGCTTTCCTCTGAGGGTCGCGCTTTAAGG

  P  L  A  Y  A  F  I  L  T  R  E  S  G  Y  P  Q  V  F  Y  G  D  M  Y  G  T  K  G  D  S  Q  R  E  I  P
  R  L  L  T  L  L  F  S  Q  G  N  L  D  T  L  R  F  S  T  G  I  C  T  G  R  K  E  T  P  S  A  K  F
  A  C  L  R  F  Y  S  H  K  G  I  W  I  P  S  G  F  L  R  G  Y  V  R  D  E  R  R  L  P  A  R  N  S

TGCCTTGAAACACAAAATTGAACCGATCTTAAAAGCGAGAAAACAGTATGCGTACGGAGCACAGCATGATTATTTCGACCACCATGACATTGTCGGCTGG
+---+---+---+---+---+---+---+---+---+---+---+---+---+---+---+---+---+---+---+---+    1500
ACGGAACTTTGTGTTTTAACTTGGCTAGAATTTTCGCTCTTTTGTCATACGCATGCCTCGTGTCGTACTAATAAAGCTGGTGGTACTGTAACAGCCGACC

   A  L  K  H  K  I  E  P  I  L  K  A  R  K  Q  Y  A  Y  G  A  Q  H  D  Y  F  D  H  H  D  I  V  G  W
  L  P  .  N  T  K  L  N  R  S  .  K  R  E  N  S  M  R  T  E  H  S  M  I  I  S  T  T  M  T  L  S  A  G
  C  L  E  T  Q  N  .  T  D  L  K  S  E  K  T  V  C  V  R  S  T  A  .  L  F  R  P  P  .  H  C  R  L

ACAAGGGAAGGCGACAGCTCGGTTGCAAATTCAGGTTTGGCGGCATTAATAACAGACGGACCCGGTGGGGCAAAGCGAATGTATGTCGGCCGGCAAAACG
+---+---+---+---+---+---+---+---+---+---+---+---+---+---+---+---+---+---+---+---+    1600
TGTTCCCTTCCGCTGTCGAGCCAACGTTTAAGTCCAAACCGCCGTAATTATTGTCTGCCTGGGCCACCCCGTTTCGCTTACATACAGCCGGCCGTTTTGC

  T  R  E  G  D  S  S  V  A  N  S  G  L  A  A  L  I  T  D  G  P  G  G  A  K  R  M  Y  V  G  R  Q  N
  Q  G  K  A  T  A  R  L  Q  I  Q  V  W  R  H  .  .  Q  T  D  P  V  G  Q  S  E  C  M  S  A  G  K  T
  D  K  G  R  R  Q  L  G  C  K  F  R  F  G  G  I  N  N  R  R  T  R  W  G  K  A  N  V  C  R  P  A  K  R

CCGGTGAGACATGGCATGACATTACCGGAAACCGTTCGGAGCCGGTTGTCATCAATTCGGAAGGCTGGGGAGAGTTTCACGTAAACGGCGGGTCGGTTTC
+---+---+---+---+---+---+---+---+---+---+---+---+---+---+---+---+---+---+---+---+    1700
GGCCACTCTGTACCGTACTGTAATGGCCTTTGGCAAGCCTCGGCCAACAGTAGTTAAGCCTTCCGACCCCTCTCAAAGTGCATTTGCCGCCCAGCCAAAG

  A  G  E  T  W  H  D  I  T  G  N  R  S  E  P  V  V  I  N  S  E  G  W  G  E  F  H  V  N  G  G  S  V  S
  P  V  R  H  G  M  T  L  P  E  T  V  R  S  R  L  S  S  I  R  K  A  G  E  S  F  T  .  T  A  G  R  F
  R  .  D  M  A  .  H  Y  R  K  P  F  G  A  G  C  H  Q  F  G  R  L  G  R  V  S  R  K  R  R  V  G  F

AATTTATGTTCAAAGATAGAAGAGCAGAGAGGACGGATTTCCTGAAGGAAATCCGTTTTTTTATTTTGCCCGTCTTATAAATTTCTTTGATTACATTTTA
+---+---+---+---+---+---+---+---+---+---+---+---+---+---+---+---+---+---+---+---+    1800
TTAAATACAAGTTTCTATCTTCTCGTCTCTCCTGCCTAAAGGACTTCCTTTAGGCAAAAAAATAAAACGGGCAGAATATTTAAAGAAACTAATGTAAAAT

   I  Y  V  Q  R  .  K  S  R  E  D  G  F  P  E  G  N  P  F  F  Y  F  A  R  L  I  N  F  F  D  Y  I  L
  Q  F  M  F  K  D  R  R  A  E  R  T  D  F  L  K  E  I  R  F  F  I  L  P  V  L  .  I  S  L  I  T  F  Y
  N  L  C  S  K  I  E  E  Q  R  G  R  I  S  .  R  K  S  V  F  L  F  C  P  S  Y  K  F  L  .  L  H  F
```

FIGURE 7.3 *(continued)*

LAB 8

PCR Amplification and Labeling
of Probe DNA

GOAL

The goal of this laboratory is to amplify a defined region of the *amyE* gene of *B. licheniformis* using the polymerase chain reaction (PCR) to produce a biotin-labeled probe.

OBJECTIVES

After completing Lab 8, you will be able to

1. outline the steps of PCR and describe what occurs during each step
2. describe applications of PCR
3. list the reagents required for PCR
4. set up PCR to amplify a specific region of DNA
5. design appropriate primers to amplify a conserved region of *Bacillus* α-amylase genes

BACKGROUND

POLYMERASE CHAIN REACTION

The polymerase chain reaction (PCR) is used to amplify (make many copies of) a region of DNA that lies between two regions of known sequence. This reaction requires two short single-stranded DNA primers that anneal (base pair) to opposite strands of the template DNA and flank the region of amplification (ROA). In addition to the template DNA and the DNA primers, two other types of biological molecules are required for this reaction: DNA polymerase and the four deoxynucleoside triphosphates (dNTPs). Amplification results from repeated cycles of the following three steps:

1. denaturation of the template DNA
2. annealing of the primers
3. extension (synthesis) of complementary strands by DNA polymerase

The three steps of PCR are shown in Figure 8.1. In step 1, heating the DNA to 95°C denatures it by breaking, or melting, the hydrogen bonds that hold the two strands together. In step 2, the temperature is lowered, and the primers hybridize,

FIGURE 8.1

The three steps of PCR cycle. ROA is the region of amplification. Note that extension of primers occurs only at the 3′ end, so synthesis is always in the 5′ to 3′ direction.

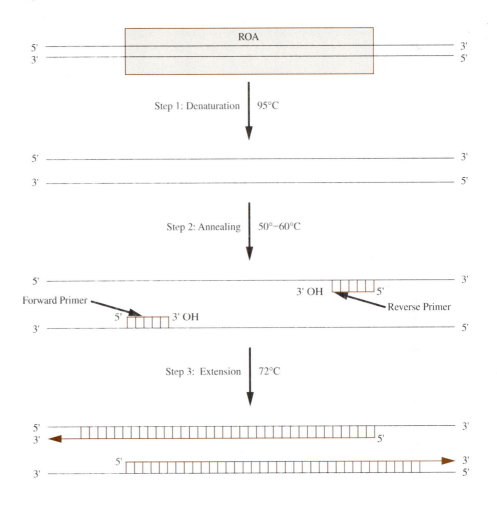

or base pair, to their complementary sequences on the template DNA. Each primer provides the free 3′-OH group needed for DNA synthesis. In step 3, the temperature is increased to 72°C, and DNA polymerase extends each primer from its 3′ end by adding deoxynucleotides in the 5′ to 3′ direction. Extension of the primers proceeds through the ROA, leading to synthesis of the complementary strand of each template strand. The polymerase used, *Taq* DNA polymerase, is from the bacterium *Thermus aquaticus,* which lives in hot springs. Because the polymerase comes from an organism that lives at high temperatures, it is not denatured at 95°C.

Since the newly synthesized strands have primer-binding sites, they can serve as template strands during subsequent cycles. During each cycle, the template strands are denatured and the primers are annealed and then extended by *Taq* DNA polymerase (Fig. 8.2). Since the products of one round of amplification serve as templates for the next, each cycle essentially doubles the number of DNA molecules. After the third cycle, the major product of this exponential amplification is a single species of double-stranded DNA whose length corresponds to the ROA plus the length of the two primers. Although longer molecules continue to be synthesized from the original template during each cycle, they accumulate at a linear rate and do not contribute much to the final mass of amplified sequences. Because of the exponential nature of PCR, 20 cycles could theoretically produce over a millionfold amplification of the target region. In practice, however, the efficiency of amplification varies, and it generally takes 25–35 cycles to make sufficient product for experimental purposes.

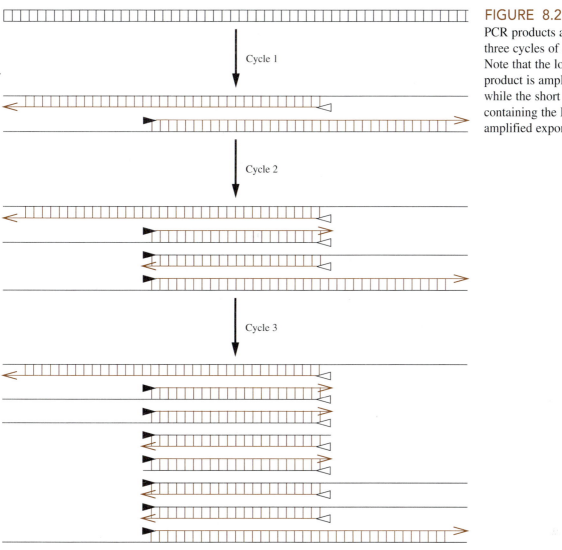

FIGURE 8.2

PCR products after the first three cycles of amplification. Note that the longer extension product is amplified linearly, while the short product, containing the ROA, is amplified exponentially.

DNA PRIMERS

PCR requires two single-stranded DNA primers. The Forward Primer is upstream of the ROA and is complementary to the lower strand (Fig. 8.3). (The sequence of the Forward Primer is identical to that of the upper strand of template DNA.) The Reverse Primer is downstream of the ROA and is complementary to the upper strand. Primers are generally 18–28 nucleotides in length and contain 50–60% guanosine and cytidine (G + C). Each primer of the pair should have a similar temperature of melting (T_m), which is defined as the temperature at which 50% of the hydrogen bonds are disrupted. Upon denaturation of the template DNA, the primers anneal to their complementary sequences, as shown in Figure 8.3. The annealing temperature is slightly below the T_m of the primers. Generally, T_m is provided by the primer supply company, but it can be estimated by multiplying the number of G + C residues by 4°C and the number of A + T residues by 2°C and adding the two numbers [$T_m = 4°(G + C) + 2°(A + T)$]. Although one could choose primers by simply analyzing the DNA sequence, computer programs can more efficiently choose primer pairs that are likely to give a good yield of product. Commercial laboratories synthesize PCR primers with the desired nucleotide sequences.

5' T T C C G G G G C G C G G C A G C A C A T A C A G C G........ROA........C A T T C A G T G T T T G A C G T G C C G C T T C A T T A T 3'

Forward Primer

Reverse Primer

3' A A G G C C C C G C G C C G T C G T G T A T G T C G C........ROA........G T A A G T C A C A A A C T G C A C G G C G A A G T A A T A 5'

FIGURE 8.3
Annealing of *amyE* primers.

DNA LABELING

Frequently it is necessary to tag, or label, DNA so that it can be detected or measured in subsequent procedures (e.g., Southern blotting). Biotin is a nonradioactive compound that is used to label DNA. It is chemically bonded to the base of a nucleotide; e.g., biotin is linked to the cytosine of dCTP (Fig. 8.4). If the PCR reaction mix contains a mixture of biotin-labeled dCTP (BIO-dCTP) and unlabeled dNTPs, then the *Taq* DNA polymerase will incorporate many molecules of BIO-deoxycytidine during PCR amplification, and the target DNA will be labeled with biotin. The biotin-labeled DNA can be detected by virtue of the strong interaction between biotin and the protein avidin. The actual detection will be provided by the enzyme alkaline phosphatase that is attached to the avidin molecule.

FIGURE 8.4
Structure of biotin-dCTP.

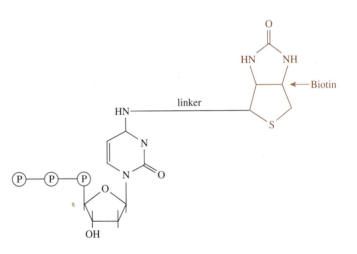

LABORATORY OVERVIEW

In this lab, you will amplify and label a region of the *amyE* gene of *B. licheniformis* with the polymerase chain reaction using the chromosomal DNA extracted in Lab 7 as template DNA. The sequences of the primers you will use to amplify a region of the *amyE* gene are shown as follows:

Forward Primer: 5'-GGGCGCGGCAGCACATAC-3'
Reverse Primer: 5'-GAAGCGGCACGTCAAACACTG-3'

The Forward Primer is located at nucleotides 700–717 of the *B. licheniformis* DNA sequence presented in Figure 7.3 and has a T_m of 67°C. The Reverse Primer is located at nucleotides 1112–1132 of the *B. licheniformis* DNA sequence presented in Figure 7.3 and has a T_m of 65°C. These two primers will amplify a PCR product of 433 bp, which includes the ROA and the two primer sequences. (The PrimerSelect program of DNAStar™ was used to design these primers.)

In the exercise at the end of this lab, you will design a pair of PCR primers that could be used to amplify a region of the *amyE* genes from several species of *Bacillus*.

TIMELINE

Diluting the DNA and setting up the reaction takes about 30–45 minutes. The reaction takes several hours to overnight, depending on your thermocycler. Gel electrophoresis takes 1.5–2.5 hours.

SAFETY GUIDELINES

Boiling agarose can cause burns. Wear protective gloves when handling hot agarose solutions.

The electric current in a gel electrophoresis device is extremely dangerous. Never remove the lid or touch the buffer once the power is turned on. Make sure that the counter where the gel is being run is dry.

Ethidium bromide is a strong mutagen. Gloves should always be worn when handling gels or buffers containing this chemical.

UV light, used to illuminate the DNA stained with ethidium bromide, is dangerous. Protect your eyes and face by wearing a UV-blocking face shield.

PROCEDURE

Your instructor has prepared two PCR mixes that must be kept on ice. (See the section on Proper Enzyme Usage in Appendix II.) The PCR Biotin Master Mix ($1\times$ PCR buffer, 2.5 mM $MgCl_2$, 50 μM dCTP, 150 μM BIO-dCTP, 200 μM dATP, 200 μM dGTP, 200 μM dTTP, 0.25 μM Forward Primer, 0.25 μM Reverse Primer, 0.05 Units/μl of *Taq* DNA polymerase) is for your experimental sample. The PCR Control Mix ($1 \times$ PCR buffer, 2.5 mM $MgCl_2$, 200 μM dNTPs, 0.05 Units/μl of *Taq* DNA polymerase) is for your control samples. ***MAKE SURE YOU USE THE CORRECT PCR MIX FOR EACH OF YOUR SAMPLES.***

ASSEMBLE THE EXPERIMENTAL REACTION

Each group will run one experimental reaction in which a portion of the *amyE* gene will be amplified and labeled with biotin. The template DNA for this reaction will be the *B. licheniformis* chromosomal DNA isolated in Lab 7. The chromosomal DNA must be diluted because PCR works best when the concentration of the template DNA is very low ($\sim$1 ng per 50 μl reaction is sufficient). Since the absolute concentration of template DNA is not critical, you will dilute your chromosomal DNA 400-fold via two (20-fold) dilutions.

   □   1.   To dilute your DNA, add 95 μl of sterile dH_2O to each of two 1.5-ml tubes. Label the first as "20-fold" and the second as "400-fold." Add 5 μl of chromosomal DNA to the first tube. Pipette the DNA slowly and carefully; check to see that the pipette tip fills properly. Vortex to mix. Add 5 μl of this dilution to the "400-fold" dilution tube. Vortex to mix and centrifuge briefly.

   □   2.   To assemble your experimental reaction, you will mix some of the 400-fold diluted DNA with the PCR Biotin Master Mix. Label a 0.5-ml PCR tube (e.g., BIO-PCR, the date, and your group number). Combine the following reagents in the tube, in the order listed. Use a new pipette tip for each reagent. Place the pipette tip against the wall of the PCR tube to ensure that the entire volume is expelled. Store your tube of 400-fold diluted DNA in the freezer; you will use it again in Lab 11A.

PCR Biotin Master Mix	49 μl
Diluted *B. licheniformis* DNA	1 μl
Total volume	50 μl

Enzyme mixtures should be handled gently. Rather than vortexing the reaction mixture, hold the tube in one hand and flick it with the other hand to mix the reagents. Centrifuge the reaction tube briefly to pull the entire sample to the bottom of the tube. (If you do not have the appropriate rotor for 0.5-tubes, place these tubes in empty 1.5-ml tubes before centrifuging.)

3. If your thermocycler does not have a heated lid, you must add ~30 μl of sterile mineral oil to the reaction tube. (Oil on the surface of the reaction mix prevents evaporation during the repeated heating and cooling cycles of PCR.) Centrifuge briefly. Keep this tube on ice until all samples are ready to be loaded into the thermocycler.

ASSEMBLE THE CONTROL REACTIONS

Two groups will work together to prepare a set of control reactions. The positive control will confirm the expected size of your experimental PCR sample. Single primers are used in negative control reactions because one primer may serve as both a Forward and Reverse Primer if there is sequence similarity elsewhere in the DNA. These controls will help you distinguish between anomalous bands and the desired product. A negative control without added DNA is used to check that none of the PCR reagents are contaminated with DNA.

1. Label a 0.5-ml PCR tube for each of the control conditions shown in Table 8.1. The template DNA for the control samples is the 433 bp fragment of the *amyE* gene, previously amplified by your instructor.
2. Assemble the control reactions by adding the reagents listed in Table 8.1 to the appropriate tube. Add the DNA last. Flick each tube to mix.
3. If necessary, add ~30 μl of sterile mineral oil to each tube. Centrifuge the tubes briefly and keep on ice until ready to load in the thermocycler.

TABLE 8.1
Components of the PCR Control Samples

Condition	PCR Control Mix	Forward Primer	Reverse Primer	dH$_2$O	Template DNA
Positive control	17 μl	1 μl	1 μl	—	1 μl
Negative control (Forward Primer only)	17 μl	1 μl	—	1 μl	1 μl
Negative control (Reverse Primer only)	17 μl	—	1 μl	1 μl	1 μl
Negative control (no DNA)	17 μl	1 μl	1 μl	1 μl	—

PERFORM PCR

☐ 1. Follow the manufacturer's directions for programming the thermocycler. The steps of this PCR program include the following:

Initial Denaturation: 95°C for 5 min
Cycle (30 times):
   Denature   95°C for 30 sec
   Anneal   62°C for 1 min
   Extend   72°C for 2 min
Final Extension: 72°C for 10 min
Hold: 12°C, 24 hr

☐ 2. Place the samples in the heating block of the thermocycler, close the lid securely, and start the program.

☐ 3. When the PCR program is complete, remove the tubes from the thermocycler. If the agarose gel will not be run in the same lab session, store the samples in the freezer until the next lab period.

VERIFY PCR PRODUCTS BY GEL ELECTROPHORESIS

☐ 1. Cast a 1% agarose gel as described in the section on Agarose Gel Electrophoresis in Appendix II.

☐ 2. If the PCR samples were frozen, place the tubes at room temperature and allow the samples to thaw completely. Centrifuge the tubes briefly. You will prepare your control reaction samples differently than your experimental, BIO-labeled PCR sample that is needed for future labs.

☐ 3. First, work with the other group and prepare your control samples. If you added mineral oil to the samples, add 20 μl of 2× DNA Gel Loading Buffer to each tube, vortex to mix, and centrifuge briefly. Carefully place the tip of an adjustable pipette below the clear layer of oil into the blue sample and withdraw as much of the sample as possible. Transfer each (blue) sample to a new, labeled 1.5-ml tube. Do not transfer any of the oil. If the oil and sample get mixed, centrifuge the tube to separate the layers.

    If you did not overlay your sample with oil, add 20 μl of 2× DNA Gel Loading Buffer to each tube, vortex to mix, and centrifuge briefly.

☐ 4. Next, prepare your experimental BIO-labeled PCR sample. If this sample was overlaid with oil, carefully withdraw the PCR reaction mixture from beneath the oil layer and transfer it to a new, labeled 1.5-ml tube. This separation may be more difficult because there is no blue dye to help you distinguish the aqueous PCR sample from the oil layer. Do not transfer any of the oil. If the oil and sample get mixed, centrifuge the tube to separate the layers. Don't worry if you cannot remove the entire PCR sample; you will recover the rest of it in the next step.

    Add 50 μl of sterile dH$_2$O to the original experimental PCR tube. Vortex to mix, and centrifuge briefly to separate the layers. Transfer the lower aqueous layer from the original PCR tube to the 1.5-ml tube containing the experimental PCR sample. Recover as much of the lower aqueous layer as possible, but do not transfer any of the oil. Vortex the 1.5-ml tube that now contains your BIO-labeled PCR sample (volume equals 100 μl), and centrifuge briefly.

    If your experimental PCR sample was not overlaid with oil, transfer the sample to a new, labeled 1.5-ml tube, and add 50 μl of sterile dH$_2$O to that tube (the final volume of BIO-labeled PCR sample equals 100 μl). (It is a good idea to transfer your sample to a larger tube because it has to be heated in future labs, and 1.5-ml tubes fit in most dry block heaters better than 0.5-ml tubes.)

☐ 5. Transfer 5 μl of your experimental BIO-labeled PCR sample to a new 1.5-ml tube. Add 5 μl of 2× DNA Gel Loading Buffer to this new tube. Vortex to mix, and centrifuge briefly. Store the remainder (~95 μl) of your experimental BIO-labeled PCR sample at −20°C for use in Lab 9C and Lab 11E.

☐ 6. Load the entire 10 μl of your experimental mixture (from step 5) into one well of a gel. Each group should load 10 μl of the positive control mixture (from step 3) in a lane next to their experimental PCR sample. Each group should load 10 μl of each of the three negative control mixtures (from step 3) on their gel. Lastly, load DNA molecular size markers (e.g., 1 kb DNA Ladder). Record in your notebook the order in which you loaded the gel.

☐ 7. Run the gel at ~90 volts until the blue dye has migrated a little more than halfway down the gel. (The PCR product migrates within the blue dye.) Stain the gel with ethidium bromide as described in Appendix II. Photograph the gel. **WEAR GLOVES WHEN HANDLING ETHIDIUM BROMIDE SOLUTIONS, AND WEAR EYE PROTECTION WHEN USING UV LIGHT.** Label your photo and place it in your lab notebook.

DATA ANALYSIS

There should be a band of ~433 bp in the lane containing the positive control sample. This band migrates slightly faster than the 500 bp marker band of the 1 kb DNA Ladder. There should be a similarly sized band in the lane with your experimental BIO-labeled PCR product. Although the intensity of staining or the thickness of the two bands may differ, they should have migrated the same distance through the gel (because they are of the same molecular size). Each lane may also have a faint, diffuse band near the bottom of the gel; these bands correspond to the DNA primers.

If your PCR product is not of the expected size, if you have more than one product, or if you have no product, discuss the possible reasons. Discuss the results of the three negative control reactions.

QUESTIONS

1. Explain the process of PCR. Include the four different biological molecules necessary and the three basic steps of this process, describing what occurs in each step.
2. Explain how biotin gets incorporated into the PCR product.
3. For what will this biotin-labeled PCR product be used?
4. High temperature normally denatures proteins so they are no longer functional. Why isn't the DNA polymerase used for PCR denatured at the temperature required to denature DNA?
5. Explain why each of the different controls was used.

EXERCISE

Generally, DNA analysis software programs (see the section on Bioinformatics in Appendix I) are used to design primers for PCR. To help you understand the principles of PCR, you will design a pair of PCR primers "by hand."

DESIGN PCR PRIMERS

On the following pages, Figure 8.5 presents the nucleotide sequences of *amyE* genes from three strains of *Bacillus* (i.e., *B. amyloliquefaciens, B. licheniformis,* and *B. stearothermophilus*). A computer program was used to align the three sequences so that highly conserved regions are apparent. The black boxes show identical nucleotides in at least two of the three sequences. Gaps (indicated by dashes) were inserted to maximize alignment among the sequences.

Use the sequence alignments to design a pair of PCR primers for amplification of an internal fragment of a *Bacillus amyE* gene. The primers should be capable of amplifying a region from all of these strains, and the amplified region should be sufficiently similar that it could serve as a probe to identify unknown *amyE* genes from other *Bacillus* strains.

The Forward Primer should match the sequence of the top strand (shown) and will bind to the bottom strand (not shown); it will be extended rightward 5′ to 3′. The Reverse Primer should match the sequence of the lower strand (not shown) and will bind to the top strand because it is complementary to that strand. It will be extended leftward 5′ to 3′. Use the following criteria for designing your PCR primers:

a. Each primer should be 18–24 nucleotides long (the limitation is cost).

b. The 4–5 nucleotides at the 3′ end of each primer must match all three *Bacillus* sequences exactly; there can be mismatches throughout the rest of the primer. When there is a mismatch, design the primer to match two of the three sequences.

c. The melting temperature (T_m) of each primer (the dissociation temperature of the primer/template duplex) should be within the range of 52°–62°C. The T_m of each primer of the primer pair should be within several degrees of the other. [T_m can be estimated as $T_m = 4°C(G + C) + 2°C(A + T)$.]

d. The percentage of G + C within each primer should be about 50%. Make sure that the G + C content is approximately the same for each of the primers.

e. The PCR product made from the primers should be 200–1000 nucleotides long.

Provide the following information about your primers:

1. Give the location (position numbers of the first and last nucleotides in the sequence) and the sequence (written 5′ to 3′) for each primer you have chosen. Remember that the sequence for the Reverse Primer is the complement of the sequence shown.
   Forward Primer: _____
   Reverse Primer: _____

2. Give the approximate T_m of each primer.
   Forward Primer: _____
   Reverse Primer: _____

3. Give the % G + C for each primer.
   Forward Primer: _____
   Reverse Primer: _____

4. What is the total size (number of base pairs including the primers) of the expected PCR product? _____

FIGURE 8.5

Alignment of *amyE* gene sequences.

LAB 9

Southern Hybridization

GOAL

The goal of this laboratory is to identify a restriction fragment from the chromosome of *B. licheniformis* that contains the *amyE* gene.

OBJECTIVES

After completing Lab 9, you will be able to

1. explain where genes are located in an organism
2. explain the action of restriction endonucleases
3. explain how an available gene, such as that for α-amylase, can be used to identify a similar gene in total chromosomal DNA

INTRODUCTION

Southern blotting and hybridization (named for E. M. Southern, who developed this technique in 1975) is a means of identifying DNA fragments that contain a particular gene or sequence of interest. In this procedure, DNA is cleaved into fragments with one or more restriction endonucleases, and the resultant fragments are separated by agarose gel electrophoresis. The gel is soaked in a denaturation solution so that double-stranded DNA is converted to single-stranded DNA. The DNA is transferred by capillary action from the gel to a membrane, to which the DNA binds tightly. The membrane is incubated with a DNA probe that is complementary to the region of interest. The single-stranded probe will hybridize, or base pair, with only complementary single-stranded DNA on the membrane. The probe DNA may contain a radioactive label or a nonradioactive label (such as biotin) that allows visualization of the hybridized fragments.

GENERAL LABORATORY OVERVIEW

This lab is divided into three sections:

Lab 9A: Chromosomal DNA from *B. licheniformis* (from Lab 7) will be cleaved with several different restriction endonucleases, and the fragments of DNA will be separated on an agarose gel.

Lab 9B: The DNA in the gel will be denatured and transferred to a membrane.

Lab 9C: The single-stranded chromosomal DNA on the membrane will be hybridized with the single-stranded, PCR-amplified, BIO-labeled *amyE* gene probe (from Lab 8). The unbound probe will be washed away, and the BIO-labeled DNA hybrids will be detected by their interaction with avidin, to which alkaline phosphatase has been linked.

TIMELINE

Day 1	Lab 9A	Set up restriction cleavages of chromosomal DNA (0.5–1 hour)
Day 2	Lab 9A	Run fragments of chromosomal DNA on a gel (2–2.5 hours)
	Lab 9B	Denature the DNA and set up a capillary transfer of the DNA to a membrane (1–1.5 hours)
Day 3	Lab 9B	Affix the DNA to the membrane
	Lab 9C	Set up the Southern hybridization (1.5–2 hours)
Day 4	Lab 9C	Detect the hybridized DNA (2.5–3.5 hours)

LAB 9A

Restriction Enzyme Cleavage of Chromosomal DNA

BACKGROUND

RESTRICTION ENDONUCLEASES

In order to identify a fragment of *Bacillus* DNA that contains the *amyE* gene, the high molecular weight chromosomal DNA must be cut into small fragments. Restriction endonucleases are the tools for this job; they are molecular scissors that cut DNA at precise locations. These enzymes bind to a specific sequence of DNA and cleave the phosphodiester bonds between the nucleotides within or near that particular sequence.

Restriction enzymes are produced by bacteria as a defense mechanism to digest the DNA of foreign invaders such as bacterial viruses (i.e., bacteriophages). Over 2000 different restriction enzymes have been discovered. A restriction enzyme is named for the genus (first letter) and species (second and third letter) of the organism from which it was isolated. The strain designation is given next in the name, followed by a Roman numeral to signify the order of restriction enzymes found in the strain. *Eco*R I, for example, is the first endonuclease isolated from the RY 13 strain of *E. coli.* The names and sources of other restriction enzyme are presented in Table 9A.1.

Most restriction endonucleases recognize and bind to a specific base sequence, called the *recognition sequence,* which is four to eight bases long and comprises inverted repeats. In an inverted repeat, or palindrome, the sequence of the top strand of DNA is identical to that of the bottom strand read in the opposite direction. Each restriction enzyme cleaves the DNA phosphodiester backbone consistently between the same two nucleotides within (or near) that sequence. The location of the cleavage sites within the recognition sequence differs from enzyme to enzyme, however. Some endonucleases, such as *Alu* I, cleave in the center of the recognition sequence of the DNA, generating fragments of DNA with blunt ends. Most others, such as *Bam*H I, *Eco*R I, *Hin*d III, and *Sau*3A I, cut each strand at a precise distance from the 5′ end of the recognition sequence, producing restriction fragments with overhanging single-stranded ends. These protruding ends are frequently called "sticky" ends because they are complementary and will base pair, or stick to, each other.

Since a given restriction enzyme always recognizes the same sequence and always generates the same ends, DNA from two different sources can be combined if they are cleaved with the same restriction enzyme. In reality, the only requirement for creating recombinant molecules is that the protruding single-stranded ends be compatible. Different enzymes can produce identical overhanging ends.

TABLE 9A.1 Examples of Restriction Enzymes			
Restriction Enzyme	**Bacterial Source**	**Recognition Sequence***	**Resulting Ends**
Alu I	*Arthrobacter luteus*	AGCT TCGA	AG CT TC GA
*Bam*H I	*Bacillus amyloliquefaciens* H	GGATCC CCTAGG	G GATCC CCTAG G
*Eco*R I	*Escherichia coli,* strain RY 13	GAATTC CTTAAG	G AATTC CTTAA G
*Eco*R V	*Escherichia coli,* strain RY 13	GATATC CTATAG	GAT ATC CTA TAG
*Hin*d III	*Haemophilus influenzae* R_d	AAGCTT TTCGAA	A AGCTT TTCGA A
*Sau*3A I	*Staphylococcus aureus* 3A	NGATCN NCTAGN	N GATCN NCTAG N

*N = any nucleotide.

For example, *Bam*H I and *Sau*3A I recognize different sequences (see Table 9A.1), but generate fragments with complementary ends that can base pair with each other. Restriction enzymes that generate fragments with blunt ends are also useful in molecular cloning. Blunt ends created by *Eco*R V, for example, can be joined not only to other blunt ends created by *Eco*R V, but to blunt ends created by any other restriction enzyme.

ENZYME REACTION CONDITIONS

The DNA to be cleaved is incubated with the desired enzyme under optimal conditions (i.e., buffer, salt, pH, and temperature) that vary widely for different restriction enzymes. Commercial manufacturers generally supply the appropriate buffer with each restriction enzyme. These buffers are supplied as concentrated solutions (e.g., 10×) and are diluted to the appropriate concentration (generally, 1×) in the reaction mixture. One unit of restriction endonuclease activity is defined as the amount of enzyme that will cleave 1 μg of DNA in 1 hour at 37°C. Restriction enzymes generally are supplied at concentrations of 10–20 units per μl.

LABORATORY OVERVIEW

The *B. licheniformis* chromosomal DNA isolated in Lab 7 will be cleaved into fragments by several different restriction enzymes. The goal is to identify a fragment that is large enough to be likely to contain the entire *amyE* gene, but small enough to clone easily into a plasmid. Each group will set up three separate reactions with the following restriction endonucleases.

> *Cla* I
> *Eco*R V
> *Hin*d III

Each enzyme will cleave the chromosomal DNA into thousands of fragments ranging in size from 0.2 kb to 20 kb. When separated on an agarose gel, these thousands of fragments of DNA will appear as a smear. The DNA in this gel will be denatured and blotted to a membrane in Lab 9B and hybridized with the BIO-labeled *amyE* gene probe in Lab 9C.

In the exercises at the end of this lab, you will locate the recognition sequences of several restriction enzymes in a short DNA sequence and answer questions about the *amyE* gene of *B. licheniformis* using information provided about the restriction cleavage sites in that sequence.

SAFETY GUIDELINES

Boiling agarose can cause burns. Wear protective gloves when handling hot agarose solutions.

The electric current in a gel electrophoresis device is extremely dangerous. Never remove the lid or touch the buffer once the power is turned on. Make sure that the counter where the gel is being run is dry.

Ethidium bromide is a strong mutagen. Gloves should always be worn when handling gels or buffers containing this chemical.

UV light, used to illuminate the DNA dyed with ethidium bromide, is dangerous. Protect your eyes and skin by wearing a UV-blocking face shield.

PROCEDURE

ASSEMBLE THE CLEAVAGE REACTIONS

☐ 1. First you must determine the amount of each reagent needed for your cleavage reactions. The final volume of each reaction will be 20 μl. You want to add about 4–6 μg of chromosomal DNA to each reaction. The amount you can add will depend upon the concentration of your sample of chromosomal DNA. Since concentration (C) × volume (V) = mass, then mass ÷ C = V. For example,

$$6\,\mu g \div 0.4\,\mu g/\mu l = 6\,\mu g \times 1\,\mu l/0.4\,\mu g = 15\,\mu l$$

$$4\,\mu g \div 0.25\,\mu g/\mu l = 4\,\mu g \times 1\,\mu l/0.25\,\mu g = 16\,\mu l$$

Determine the volume of 10× stock buffer to add to each 20 μl reaction to get a final concentration of 1×. Calculate as follows:

$$V_1 C_1 = V_2 C_2$$

$$?\,\mu l\,(10\times) = 20\,\mu l\,(1\times)$$

$$?\,\mu l = 20/10 = 2\,\mu l$$

In general, 1.0 μl of restriction enzyme contains sufficient enzyme (10–20 units) to digest this amount of DNA. Calculate the volume of dH_2O needed to bring the total volume of the reaction to 20 μl. Fill in the blanks that follow and record this in your notebook.

dH_2O	_____ μl
B. licheniformis DNA	_____ μl (containing 4–6 μg of DNA)
10× enzyme buffer	2.0 μl (1× final concentration)
Restriction enzyme	1.0 μl
Total volume	20.0 μl

2. Completely thaw your tube of *B. licheniformis* DNA from Lab 7. Flick the tube to mix, and centrifuge briefly. Completely thaw the 10× restriction enzyme buffers (provided by the manufacturer of the enzymes you are using). Do not use buffers that are partially thawed, as the salt and buffer concentrations will be incorrect. Vortex to mix, centrifuge briefly, and store on ice. (Refer to the section on Proper Enzyme Usage in Appendix II.) Briefly centrifuge the tubes of restriction enzymes. Keep the enzymes on ice at all times to minimize inactivation.

3. Label a sterile, 1.5-ml microcentrifuge tube for each of the three cleavage reactions and include your group number or initials. Based on your calculations in step 1, add the appropriate volume of each reagent, in the order listed, to each labeled microcentrifuge tube. Use a new tip for each reagent. Use the appropriate buffer for each enzyme. *Note:* The enzymes are provided in a glycerol storage solution; to measure accurately, you must pipette slowly and carefully.

4. Prepare a control reaction of undigested DNA. This tube should contain about 1 μg of chromosomal DNA, buffer, and water, but no restriction enzyme. Calculate the volume of chromosomal DNA to be added to the tube. Adjust the volume of dH$_2$O to provide a total volume of 20 μl.

5. Flick each tube to mix, and centrifuge briefly. Incubate all the tubes at 37°C for at least 4 hours. You or your instructor will remove the digestions from the incubator and store them in the freezer until the next lab session.

SEPARATE THE DNA FRAGMENTS BY GEL ELECTROPHORESIS

1. Cast a 0.7% agarose gel according to the directions in Appendix II. Make sure that the wells of the gel will hold ~25 μl.

2. Remove the restriction enzyme cleavage reactions and the uncut control DNA from the freezer and thaw completely. Spin briefly to pull any condensation into the bottom of each tube.

3. Add 3 μl of 6× DNA Gel Loading Buffer (0.25% Bromophenol Blue, 50 mM EDTA, 40% sucrose) to the 20 μl in each tube. Vortex to mix, and centrifuge briefly.

4. Load the entire volume of each of the restriction cleavage reactions and the undigested control DNA in adjacent lanes of the gel. Write down (in your notebook) the order in which you loaded the samples. It is a good idea to load the samples in alphabetical order so that you can readily recall the order of loading.

5. You also need to load DNA molecular size markers and a positive control (i.e., biotinylated DNA) for the Southern hybridization (Lab 9C). You can accomplish both of these by using biotinylated DNA size markers, e.g., the bacteriophage lambda (λ) DNA digested with *Hin*d III and labeled with biotin (biotinylated λ DNA/*Hin*d III). These markers can also be visualized with ethidium bromide staining. This marker should be heated to 65°C for 5 minutes before loading on the gel. Your instructor will tell you how much to load and will provide you with information about these marker bands. You can also load unlabeled DNA markers (e.g., 1 kb DNA Ladder), if desired. Keep a record of which lanes were loaded with the size markers.

6. To save time, you may add ethidium bromide directly to the running buffer (refer to the section on Staining Gels in Appendix II). Run at

80–90 volts for 1.5–2 hours or until the blue dye is about three-fourths of the way down the gel.

☐ 7. ***WEAR GLOVES WHEN HANDLING ETHIDIUM BROMIDE SOLUTIONS, AND WEAR EYE PROTECTION WHEN USING UV LIGHT.*** Carefully remove the gel from the chamber. Cut off the upper right corner of the gel, above the wells, so that you will know when the gel is right side up. Photograph the gel; label and place the photo of the gel in your notebook. ***DO NOT DISCARD THE GEL,*** but proceed immediately to Lab 9B.

DATA ANALYSIS

- Determine whether your chromosomal DNA was digested by each of the three restriction enzymes. A smear of DNA along the length of the gel with no discernible band of high molecular weight DNA close to the wells indicates that the DNA was cut into thousands of fragments. There may be discrete bands visible within the smear of fragments in some of the lanes. (There may be a bright smear of RNA near the bottom of each lane.)
- Compare the chromosomal DNA that was not digested with a restriction enzyme with the DNA that was digested. Does the undigested DNA appear as a single, high molecular weight band? If your undigested DNA appears degraded (i.e., as a smear), you should not use it for the cloning experiment in Lab 10. Borrow some DNA from a group whose undigested DNA does not appear to be degraded.
- Determine the range of fragment sizes produced by each of the enzymes. What can you conclude about frequency of recognition sequences for an enzyme that generates fragments ranging in size from 5–20 kb versus an enzyme that generates fragments ranging in size from 0.5–6 kb?

EXERCISES

Once the nucleotide sequence of a piece of DNA is known, the location of restriction enzyme cleavage sites can be determined. Generally, computer software programs are used to search DNA sequences and find the recognition sequences of restriction endonucleases. (See the Bioinformatics section and the Sequence Analysis Using Computer Software exercise in Appendix I.) In the first exercise that follows, you will manually search a short sequence of DNA for restriction enzyme cleavage sites. In the second exercise, you will analyze a computer-generated restriction map of the *amyE* gene of *B. licheniformis*.

EXERCISE 9.1
Locate Restriction Enzyme Cleavage Sites

Analyze the following DNA sequence and determine whether it contains any recognition sequences for the restriction enzymes listed in Table 9A.2. Record in the table the number of sites and the position of the site of cleavage in this DNA sequence.

```
        1           5          10          15          20          25
5'- T A A A G C C G G G C A A A T G A A A A T G G A T C -3'
```

Restriction Enzyme	Recognition Sequence[1] and Site of Cleavage[2]	Number of Sites in Sequence Provided	Position of Cleavage Site in Sequence Provided[3]
Dpn I	GA↓TC		
Dsa V	↓CCNGG		
Dra I	TTT↓AAA		
Hpa II	C↓CGG		
Mbo I	↓GATC		
ScrF I	CC↓NGG		
Taq I	T↓CGA		

TABLE 9A.2
Identification of Recognition Sequences

[1]N = any nucleotide.

[2]The site of cleavage is indicated by the arrow in the recognition sequence.

[3]The position of the cleavage site refers to the nucleotide before the cleavage site in the recognition sequence.

EXERCISE 9.2
Restriction Analysis of the *amyE* Gene

A map showing the location of restriction enzyme cleavage sites in a fragment of *B. licheniformis* DNA that contains the *amyE* gene is shown in Figure 9A.1. This is a physical map of the DNA sequence that was presented on pp. 83–84; it was generated by a DNA analysis software program. The numbers along the top of the diagram designate nucleotide positions in the sequence. The arrows at the bottom

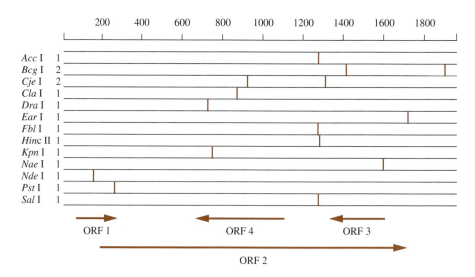

FIGURE 9A.1
Restriction map of *B. licheniformis amyE* gene region. The map indicates the cleavage sites of several restriction enzymes in a fragment of *B. licheniformis* DNA that contains the α-amylase gene.

of the diagram show the length and direction of several of the open reading frames (ORFs) in this region of DNA. The number to the right of each enzyme's name indicates the number of restriction enzyme cleavage sites in this sequence. The vertical bars on each line indicate the location of the cleavage sites of that enzyme. Use this restriction map to answer the following questions.

1. Which ORF shown along the bottom of the diagram represents the correct one for the *amyE* gene? _____
   _____

2. If you wanted to cleave this DNA with a single restriction enzyme that would give one or more fragments entirely inside the coding region of the gene, which enzyme or enzymes would you use? _____
   _____

3. What would be the approximate size of the DNA fragment or fragments that result from cleavage with the enzyme(s) you chose in question 2 (see the nucleotide scale at the top)? _____
   _____

4. Which two enzymes would you choose if you wanted to isolate a DNA fragment that contained *most* of the *amyE* gene, but *no* DNA outside of the coding region? _____
   _____

5. What would be the approximate size of the DNA fragment that results from cleavage with the enzymes you chose in question 4 (see scale at the top)? _____
   _____

6. One method for creating a mutation in a cloned gene is to cleave the DNA with an enzyme that cuts twice within the gene of interest and then join (ligate) the ends (which are compatible with each other because they were made by the same enzyme). The resulting DNA has a deletion in the gene of interest (i.e., the region between the two original restriction cleavage sites is deleted). Which enzyme could be used to create a deletion within the coding region of the *amyE* gene? _____
   _____

LAB 9B

Denaturation and Transfer of DNA to a Membrane

BACKGROUND

The separated fragments of DNA in the agarose gel must be denatured, or made single-stranded, so that the single-stranded DNA probe can base pair with its complementary DNA during the hybridization step. To denature the DNA, the gel is soaked in a strong alkaline solution, which breaks the hydrogen bonds between the two strands of the DNA. If the fragments of interest are larger than 15 kb, transfer to the membrane may be improved if the DNA is depurinated by acid treatment prior to alkaline denaturation.

If the DNA is to be transferred to a nitrocellulose membrane, the gel must be neutralized before transfer. However, if the DNA is to be transferred to a nylon membrane, the transfer can proceed under alkaline conditions. The denatured DNA is transferred to the membrane by capillary action. The membrane is placed directly on the gel, and absorbent paper is placed on top of the membrane to draw the transfer solution from the gel upwards. The DNA will thus be pulled out of the gel and onto the membrane. Since the DNA cannot pass through the membrane, it remains bound to the membrane.

The DNA must be permanently bound to the membrane either by baking the membrane until it is completely dry or by exposing it to UV light at 254 nm. The mechanism by which DNA binds permanently to a membrane by baking is not completely understood. The binding may be due to hydrogen bonding because single-stranded, but not double-stranded, DNA will bind to some membranes. UV light leads to the formation of covalent bonds between the thymine bases of DNA and the membrane.

LABORATORY OVERVIEW

You will denature the DNA by incubating the gel in a salt solution containing 0.4 M sodium hydroxide. After a brief soak in an alkaline transfer solution, the DNA in the gel will be transferred to a nylon membrane. The Southern blot setup is a modification of the traditional setup in that it uses only the liquid in the gel for transfer.

SAFETY GUIDELINES

Ethidium bromide is a strong mutagen and a possible carcinogen. Gloves should always be worn when handling gels or buffers containing this chemical.

UV light, used to illuminate the DNA stained with ethidium bromide, is dangerous. Protect your eyes and face by wearing a UV-blocking face shield.

The Southern Denaturation Solution contains 0.4 M sodium hydroxide, which is mildly corrosive at this concentration. Wear gloves when handling this solution.

Gloves should always be worn when handling transfer membranes since the oils and proteins on your bare hands can interfere with the binding of DNA and can lead to elevated background staining.

PROCEDURE

DENATURE THE DNA

☐ 1. ***WEAR GLOVES WHEN HANDLING ETHIDIUM BROMIDE SOLUTIONS, AND WEAR EYE PROTECTION WHEN USING UV LIGHT.*** Place your ethidium bromide-stained gel on the transilluminator and trim away any unused lanes. Cut off the upper right corner of the gel, if necessary.

   If you did not use biotinylated DNA molecular size markers, you must make an overlay of the gel so that you will be able to estimate the sizes of the hybridized fragments at the end of this lab. (Even if you did use biotinylated DNA markers, it is still a good idea to make an overlay of the gel.) Place a piece of clear plastic wrap over the stained gel, smoothing out any wrinkles. Mark the edges of the gel, the wells, and the DNA molecular size marker bands using a fine-tipped marking pen (it is not necessary to mark the smears of digested DNA). Since this overlay will be placed over the finished blot to determine the size of the DNA bands that contain the *amyE* gene, it is very important that the wells and molecular size marker bands be marked accurately. Remove the piece of plastic wrap, rinse the side that touched the gel with water, and blot dry with a paper towel. This overlay can be traced for the other member(s) of the group. Put your overlay in your notebook.

☐ 2. Place the gel into a shallow container. Add ~100 ml of Southern Denaturation Solution (3 M NaCl, 0.4 M NaOH) to denature the DNA. Place the container on a rocker or shaker set at low speed and shake gently for 30 minutes.

☐ 3. Remove the container from the rocker and carefully pour off and discard the solution while holding the gel in the container with a gloved hand.

> �֎ CAUTION: The denaturation solution makes the gel extremely slippery and fragile!

☐ 4. Add ~100 ml of Southern Transfer Solution (3 M NaCl, 8 mM NaOH) and gently shake or rock the gel for 15 minutes. While waiting, cut your transfer membrane and absorbent paper (see steps 1 and 4 that follow).

TRANSFER THE DNA TO A MEMBRANE

☐ 1. ***WEAR GLOVES WHEN HANDLING TRANSFER MEMBRANES.*** Cut a piece of nylon membrane several millimeters larger than your gel. *Note:* In the package, the membrane is generally sandwiched between two pieces of protective paper. Write your group number or initials near one end of the membrane with a soft lead

FIGURE 9B.1
Southern transfer setup.

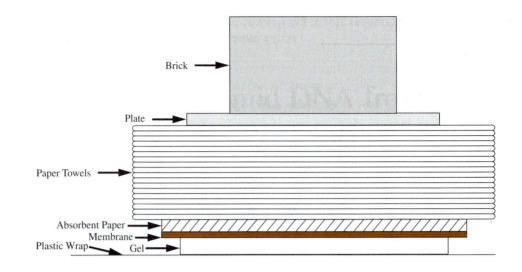

pencil; do not use ink because it will run during the hybridization procedure. Wet the membrane by placing it in a shallow dish of dH₂O. When the entire membrane is wet, replace the water with ~20 ml of Southern Transfer Solution, submerge the membrane, and let it soak for several minutes.

2. Set up the Southern transfer, as diagrammed in Figure 9B.1. Place a piece of plastic wrap on the counter. Carefully place the gel, right side up, on the plastic wrap. Remember, the upper right-hand corner above the wells was cut off.

3. Place the wet transfer membrane directly on top of the gel, with your pencil marks facing the gel. Use a glass pipette in a rolling-pin fashion to remove any air bubbles trapped between the gel and the membrane.

4. Place a piece of thick absorbent paper (e.g., Whatman 3 MM paper), cut slightly larger than the membrane, on top of the transfer membrane.

5. Place a stack of paper towels, ~1 inch thick, on top of the absorbent paper.

6. Place a glass or plastic plate on top of the paper towels. This plate helps to distribute the weight of the brick.

7. Place a brick, a book, or other weight on top of the plate. Allow the transfer to proceed for 12–24 hours.

8. To disassemble the blot, discard the paper towels and absorbent paper. Turn over the gel and transfer membrane so that the gel faces up. Use a soft lead pencil to mark the position of each well onto the membrane (press the tip of the pencil through the thin layer of gel). You can also mark the edges of the gel on the membrane; this will help you align the blot and overlay at the end of the lab. Discard the gel. The DNA is on the side of the membrane that faced the gel and has the pencil marks. (Your blot will be a mirror image of your gel.)

9. Neutralize the membrane by soaking it in ~20–25 ml of Membrane Neutralization Solution (0.2 M sodium phosphate, pH 6.8) for 3 minutes. Air-dry the membrane and wrap it in foil. (The blue dye on the membrane will disappear during the subsequent hybridization and rinsing steps of Lab 9C.)

AFFIX THE DNA TO THE MEMBRANE

If the DNA will be linked to the membrane with UV light, choose the setting suggested by the manufacturer of the UV cross-linker. Make sure that the DNA side of the blot faces up so that it is fully exposed to the UV light.

If the DNA will be bound to the membrane by baking, place the membrane between two pieces of filter paper and bake it at 80°C for 30 minutes to 2 hours. A vacuum oven is required for baking nitrocellulose membranes but is not required for baking nylon membranes.

DATA ANALYSIS

- Upon completion of your Southern hybridization (Lab 9C), you will be able to determine whether your DNA was denatured, transferred, and bound to the nylon membrane.

LAB 9C

Southern Hybridization and Detection

BACKGROUND

HYBRIDIZATION

The fragments of DNA containing the gene of interest are identified by their hybridization to a single-stranded DNA probe that is complementary to the region of interest. Temperature and ionic strength of the hybridization solution affect the base pairing of single-stranded probe DNA to its complementary DNA. *Stringency* refers to the reaction conditions under which hybridization occurs. At high stringency, base pairing occurs only between strands with perfect complementarity, whereas some mismatches of bases between the strands are tolerated at lower stringency. The DNA strands readily hybridize at low temperatures and high salt concentrations, whereas higher temperatures and lower salt concentrations provide more stringent conditions. Generally, high stringency hybridizations are performed in the presence of $\sim 5 \times$ SSC (Standard Saline Citrate; 0.75 M sodium chloride, 75 mM sodium citrate), and the temperature depends upon the solvent used (i.e., $\sim 65°C$ in aqueous solution or $42°C$ in 50% formamide). The hybridization solution generally also contains blocking agents (i.e., proteins and nonspecific nucleic acids) to bind to nonspecific sites on the membrane and reduce background staining.

To prevent nonspecific adsorption of the probe DNA to the membrane, the membrane is preincubated, or prehybridized, with hybridization solution that does not contain the probe DNA. The proteins and nonspecific nucleic acids in this solution adsorb to the membrane wherever DNA is not bound.

After hybridization, the membrane is washed extensively to remove the unbound probe. The washing conditions should be as stringent as possible, i.e., at low ionic strength and higher temperatures.

DETECTION

Biotin-labeled DNA can be detected by its interaction with avidin (or streptavidin) to which an enzyme or a fluorescent molecule has been attached. If the enzyme alkaline phosphatase is linked to streptavidin, then the hybridized target DNA-biotin probe-streptavidin complex can be detected colorimetrically. As detailed in Lab 5B, alkaline phosphatase converts a soluble, colorless substrate (BCIP/NBT) into an insoluble, colored product. Thus, colored bands appear wherever the biotin-labeled probe DNA hybridized with its complementary DNA on the membrane (Fig. 9C.1).

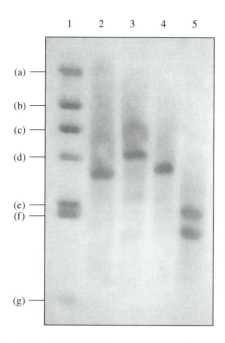

Southern hybridization. *B. licheniformis* chromosomal DNA probed with biotinylated *amyE* gene probe. Lane 1: biotinylated λ DNA/*Hin*d III molecular weight markers—(*a*) 23.1 kb, (*b*) 9.4 kb, (*c*) 6.6 kb, (*d*) 4.4 kb, (*e*) 2.3 kb, (*f*) 2.0 kb, (*g*) 0.5 kb; lane 2: DNA was cleaved with *Hin*d III; lane 3: DNA was cleaved with *Eco*R V; lane 4: DNA was cleaved with *Eco*R I; lane 5: DNA was cleaved with *Cla* I.

LABORATORY OVERVIEW

The membrane with the attached fragments of *B. licheniformis* chromosomal DNA will be incubated overnight in a hybridization solution containing the BIO-labeled *amyE* PCR probe prepared in Lab 8. Since the 433 bp PCR probe is identical to the target DNA (*amyE* gene sequence), your hybridization will be carried out at high stringency (i.e., 42°C in the presence of 5× SSC, formamide, Denhardt's reagent [a mixture of bovine serum albumin, Ficoll, and polyvinylpyrrolidone], and herring sperm DNA). After a series of high stringency washes, the restriction fragments that hybridized to the BIO-labeled *amyE* probe will be visualized by the activity of alkaline phosphatase, which is linked to streptavidin (AP-streptavidin).

Southern hybridization will allow you to determine which restriction enzyme produces a 2.5–5 kb fragment that most likely contains the entire *amyE* gene and is suitable for cloning. If the chromosomal DNA was digested completely, each restriction enzyme should produce one fragment that hybridizes to the *amyE* PCR probe. If two or more hybridizing fragments are generated by any of restriction enzymes, this enzyme cleaves the chromosomal DNA somewhere within the region of the *amyE* gene that was amplified by PCR (Fig. 9C.2). Such enzymes are not useful if you want to clone the entire gene.

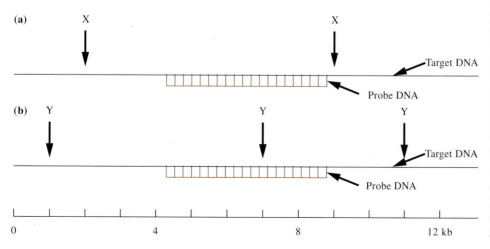

Interpreting Southern hybridization results. Recognition sequences of two different restriction endonucleases (X and Y) are shown. (*a*) The recognition sequences (arrows) of Enzyme X lie outside the region hybridized to the probe DNA; therefore, one 7-kb band hybridizes. (*b*) The recognition sequences of Enzyme Y lie outside and within the region hybridized to the probe DNA; therefore, two bands of 6 kb and 4 kb hybridize.

SAFETY GUIDELINES

Wear gloves when handling the nylon membrane to prevent the transfer of oils, proteins, and other contaminants from your hands to the membrane. Handle the membrane by the edges only and keep it as flat as possible during all the steps because creases, excess pressure, and points of contact will develop elevated background staining.

PROCEDURE

HYBRIDIZE THE MEMBRANE WITH BIO-LABELED PROBE DNA

1. ☐ ***WEAR GLOVES WHEN HANDLING NYLON MEMBRANES.*** Soak the membrane for several minutes in a shallow dish containing ~20 ml of 2× SSC (0.3 M NaCl, 30 mM Na-citrate).

2. ☐ Heat the tube of sheared herring sperm DNA at 100°C for 10 minutes to make it single stranded. Chill on ice. Add 0.25 ml of the heat-denatured, herring sperm DNA to 5 ml of the Southern Prehybridization Solution. (Upon addition of this nonspecific DNA, the Southern Prehybridization Solution contains 50% formamide, 5× SSC, 5× Denhardt's Reagent, 20 mM sodium phosphate, pH 6.5, and 0.5 mg/ml denatured, sheared herring sperm DNA.) Mix. Keep on ice until ready to use.

3. ☐ Place the wet membrane in a quart-sized, zippered freezer bag. (Use a separate bag for each membrane.) Add the Southern Prehybridization Solution (from step 2) to the bag. Remove the trapped air and air bubbles before sealing the bag. If you are using a water bath for hybridization, place the bag with the membrane into a larger freezer bag to prevent water leakage. Remove trapped air from the second bag before sealing.

4. ☐ Incubate at 42°C for at least 1 hour. Gentle agitation is helpful but not essential.

5. ☐ Near the end of the prehybridization period, heat the tube of sheared herring sperm DNA at 100°C for 10 minutes. Chill on ice. Add 100 μl of heat-denatured, sperm DNA to 5 ml of Southern Hybridization Solution. (Southern Hybridization Solution contains 45% formamide, 5× SSC, 1× Denhardt's Reagent, 20 mM sodium phosphate, pH 6.5, and 0.2 mg/ml of denatured, sheared herring sperm DNA.) Mix. Keep on ice until ready to use.

6. ☐ Heat your tube of BIO-labeled PCR probe from Lab 8 at 100°C for 10 minutes to denature the DNA. Chill on ice. Centrifuge briefly. Add 60 μl of your heat-denatured, BIO-labeled PCR probe to the tube of Southern Hybridization Solution (step 5). Mix. Keep in ice. (Store the rest of the BIO-labeled PCR probe at −20°C for use in Lab 11E.)

7. ☐ Remove your bag containing the membrane from the incubator. Discard the prehybridization solution and add the Southern Hybridization Solution + BIO-labeled probe (from step 6). Remove the trapped air and air bubbles before sealing the bag. If you are using a water bath for hybridization, place the hybridization bag into a larger freezer bag to prevent water leakage. Remove trapped air from the larger bag before sealing.

8. ☐ Incubate at 42°C for at least 16 hours. Gentle agitation is helpful but not essential.

    Note: After the incubation period, the bag containing the membrane and the hybridization solution can be stored at −20°C for several days.

DETECT THE BIO-LABELED HYBRIDIZED DNA

keep the
hybridization soln.

- [ ] 1. **WEAR GLOVES WHEN HANDLING NYLON MEMBRANES.** If the membrane and hybridization solution were stored at −20°C, you may have to briefly warm the bag to ~40°C (to bring the salts into solution). Remove the membrane from the bag and place it in a shallow container or a new, quart-sized, zippered bag. Add ~100 ml of Southern Wash Solution 1 (2× SSC, 0.1% SDS) to the membrane and rotate gently at room temperature for 3 minutes. If using a zippered bag, remove most of the air bubbles.

- [ ] 2. Transfer the used Southern Hybridization + BIO-Probe Solution (from the hybridization bag) to a round-bottom 15-ml tube. Label the tube and store at −20°C; this solution will be used again in Lab 11E.

- [ ] 3. Drain off the wash solution from the membrane. Add another 100 ml of Southern Wash Solution 1 and rotate at room temperature for 3 minutes.

- [ ] 4. Discard the wash solution and add about 100 ml of Southern Wash Solution 2 (0.2× SSC, and 0.1% SDS). Rotate gently at room temperature for 3 minutes. Discard the wash solution. Repeat one time.

- [ ] 5. Discard the wash solution and add about 100 ml of Southern Wash Solution 3 (0.16× SSC, 0.1% SDS) that has been prewarmed to 65°C. Incubate for 10 minutes at 65°C. Repeat one time.

- [ ] 6. Rinse the membrane briefly in about 50 ml of Tris-NaCl (100 mM Tris-HCl, pH 7.5, 150 mM NaCl). Discard the rinse solution and add about 30 ml of Biotin Blocking Buffer (3% bovine serum albumin in Tris-NaCl). Incubate at 65°C for 45–60 minutes.

- [ ] 7. Discard the blocking solution and transfer the membrane to a new zippered bag. Add 10 ml of AP-Streptavidin Solution (1 μg/ml AP-Streptavidin in Biotin Blocking Buffer). Incubate at room temperature for 10–15 minutes with gentle agitation.

- [ ] 8. Transfer the membrane to a shallow container and wash in about 200 ml of Tris-NaCl for 5 minutes at room temperature with gentle agitation. Discard the solution. Repeat this wash step three more times.

- [ ] 9. Wash the membrane in about 50 ml of AP Reaction Buffer (100 mM Tris-HCl, pH 9.5, 100 mM NaCl, 50 mM $MgCl_2$) for 3–5 minutes at room temperature with gentle agitation.

- [ ] 10. Completely drain the reaction buffer and add 20 ml of AP Substrate Solution (BCIP and NBT in AP Reaction Buffer) that was freshly made by your instructor. Allow the reaction to proceed in the dark or in low light. Check the membrane occasionally until blue-purple bands appear. (Membranes can be left in the substrate solution overnight but the background color will be elevated, leading to a decreased signal to noise ratio.)

- [ ] 11. Briefly rinse the membrane in about 50 ml of AP Stop Buffer (20 mM EDTA in phosphate buffered saline at pH 7.5) to terminate the color reaction.

- [ ] 12. Photocopy the wet membrane (it may help to wrap the membrane in plastic) and place the photocopy in your lab notebook. Wrap the original membrane in aluminum foil to keep light from degrading the color of the bands.

DATA ANALYSIS

- The restriction fragments of *B. licheniformis* DNA containing the *amyE* gene should be visible as blue-purple bands on the membrane. What can you conclude if the only bands visible on your blot are those of the biotinylated marker DNA? What can you conclude if no bands are visible anywhere on the blot? Why is it important to include a positive control (i.e., biotinylated DNA) on Southern blots?

- Estimate the size of each of your hybridized *amyE* gene-containing fragments. First, you must construct a standard curve, plotting molecular size versus distance migrated. Use your Southern blot or the overlay of the original gel to measure the distance migrated by each of the DNA molecular size markers. Measure from the bottom of the well to the middle of each band. Plot the size (in kb) of each DNA size marker on the y-axis and the distance it migrated on the x-axis on semilog graph paper. Draw a straight best-fit line through the data points that lie on the linear portion of the curve. (It is possible that the largest and/or smallest molecular size markers do not fall in the linear range, resulting in an S-shaped curve. If this is the case, draw your line through at least four points that clearly lie on the linear portion of the curve.) Second, measure the distance that each of your hybridized fragments migrated. Measure from the bottom of the well to the middle of the band. Use your standard curve to determine the size (in kb) of each of the hybridized restriction fragments. Organize your data into a table to place in your notebook, along with your standard curve.

- Share your data with the rest of the class. Do all groups agree on the number and size of the fragments that hybridized with the BIO-labeled *amyE* gene probe? If more than one band appears for any restriction enzyme, what does this indicate? Which of the three restriction enzymes generated an *amyE* gene-containing fragment of the appropriate size for cloning?

QUESTIONS

1. Explain the action of restriction enzymes.
2. Explain why the gel was soaked in strong alkaline solution prior to transfer to a membrane.
3. Explain at the molecular level how a probe that is made from a known fragment of DNA can be used to identify a specific gene on chromosomal DNA fragments separated by gel electrophoresis.
4. If you used your BIO-labeled *amyE* PCR probe to identify restriction fragments containing the *amyE* gene in another *Bacillus* species, would you increase or decrease the stringency of the hybridization reaction? Why?
5. Using your Southern blot, identify the restriction enzymes that did not digest within the probe region. Does this also mean that these enzymes do not digest within the *amyE* gene? Explain your answer.
6. The map that follows shows the locations of the recognition sequences (arrows) of two restriction enzymes (A and B) within a 10 kb piece of DNA. List the sizes of the fragments that would be generated upon digestion of this DNA with each of the enzymes. Indicate which of these fragments would hybridize to the probe DNA, diagrammed on the map, during Southern blotting.

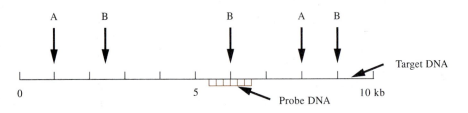

LAB 10

Cloning the α-Amylase Gene

GOAL

The goal of this laboratory is to clone the *amyE* gene from *B. licheniformis* into *E. coli.*

OBJECTIVES

After completing Lab 10, you will be able to explain and perform the following techniques:

1. restriction enzyme cleavage of DNA
2. ligation
3. transformation
4. screening for recombinant clones

INTRODUCTION

Cloning is the process by which a group of identical molecules, cells, or organisms is created. Genes can be cloned by inserting a fragment of DNA containing the gene of interest into a second piece of DNA (called the vector) that can replicate in a cell, producing many copies of the gene of interest. The cloning vector can be a bacteriophage (a virus that infects bacterial cells) or a plasmid (a small circular molecule of DNA that replicates independently of the bacterial chromosome).

To clone a chromosomal gene, the chromosomal DNA is cleaved into fragments with a restriction endonuclease that leaves the gene of interest on one fragment. (Southern hybridization is used to identify the restriction fragment containing the gene of interest.) A suitable cloning vector is chosen (e.g., a plasmid) and cut open, or linearized, with the same restriction endonuclease. The fragments of chromosomal DNA and the vector DNA are mixed together, and their ends are linked together, or ligated, by the enzyme DNA ligase. This mixture of recombinant DNA molecules is then introduced into bacterial cells by a process called *transformation,* and the resulting collection of clones is called a *DNA library.* Since the plasmid cloning vector carries an antibiotic resistance gene, cells that have been transformed with the plasmid can be selected by using an antibiotic to kill the cells that were not transformed. More sophisticated screening procedures are needed to distinguish clones with recombinant plasmids from those with non-recombinant plasmids. Screening for recombinant plasmids containing the DNA of interest can be accomplished by hybridization using the same probe that identified the restriction fragments, or, in some cases, by assaying for an enzyme produced by the gene of interest. In this laboratory, you will identify clones of *E. coli* that contain the *amyE* gene of *B. licheniformis* by their ability to digest starch.

GENERAL LABORATORY OVERVIEW

This lab is organized into five sections:

Lab 10A: *B. licheniformis* chromosomal DNA (from Lab 7) will be cleaved into fragments with *Hin*d III.

Lab 10B: A plasmid cloning vector will be linearized with *Hin*d III.

Lab 10C: The fragments of chromosomal DNA will be ligated with the cleaved plasmid DNA.

Lab 10D: *E. coli* cells will be transformed with the ligation mixture.

Lab 10E: Clones containing the *amyE* gene will be identified by their ability to degrade starch.

TIMELINE

Day 1	Lab 10A	Set up *Hin*d III cleavage reaction of chromosomal DNA (0.5 hour)
	Lab 10B	Set up *Hin*d III cleavage reaction of plasmid DNA (0.5 hour)
Day 2	Lab 10A,B	Run gel to check that DNA species have been cleaved (1.5–2 hours)
	Lab 10C	Set up ligation reaction (0.5 hour)
Day 3	Lab 10D	Transform *E. coli* cells with ligation mixture (1.5 hours) and spread transformation reaction on LB agar starch plates (0.5 hour)
Day 4	Lab 10E	Primary screening: Stain plates with Lugol Solution (1 hour), pick positive colonies, serially dilute, and spread on LB agar starch plates (0.5 hour)
Day 5	Lab 10E	Secondary screening: Stain plates with Lugol Solution (0.5 hour), pick isolated positive colonies, and streak on LB agar (stock) plates (0.5 hour)
	Lab 11A	Set up PCR reactions (0.5 hour)

LAB 10A

Cleavage of Chromosomal DNA

LABORATORY OVERVIEW

Your Southern blot analysis of Lab 9 revealed that the *amyE* gene of *B. licheniformis* is on a 3.3 kb *Hin*d III fragment. This fragment should be large enough to contain the entire (1.6 kb) *amyE* gene and its control elements. To clone this gene, you will cleave the *B. licheniformis* DNA isolated in Lab 7 with *Hin*d III. These thousands of restriction fragments will be joined with plasmid DNA to create a library of genes from *B. licheniformis*.

SAFETY GUIDELINES

Boiling agarose can cause burns. Wear protective gloves when handling hot agarose solutions.

The electric current in a gel electrophoresis device is extremely dangerous. Never remove the lid or touch the buffer once the power is turned on. Make sure that the counter where the gel is being run is dry.

Ethidium bromide is a strong mutagen and a possible carcinogen. Gloves should always be worn when handling gels or buffers containing this chemical.

UV light, used to illuminate the DNA dyed with ethidium bromide, is dangerous. Protect your eyes and face by wearing a UV-blocking face shield.

PROCEDURE

CLEAVE CHROMOSOMAL DNA WITH *HIN*D III

☐ 1. Determine the amount of each reagent to add to your reaction tube; the final volume of the reaction will be 20 μl. You want to add about 0.5 μg of chromosomal DNA. Calculate the volume of chromosomal DNA to add (see the sample calculations on p. 101, if necessary). Calculate the volume of dH$_2$O needed to bring the total volume of the reaction to 20 μl (1 μl of enzyme will be more than enough for this amount of DNA). Fill in the blanks that follow and record this in your notebook.

dH$_2$O	__ μl
B. licheniformis DNA	__ μl (containing 0.5 μg of DNA)
10× *Hin*d III buffer	2 μl (1× final concentration)
*Hin*d III	1 μl
Total volume	20 μl

Handwritten in margin:
9.5
7.5
2
1
—
20.0

☐   2. Completely thaw the tube of *B. licheniformis* chromosomal DNA from Lab 7. Flick the tube to mix, and centrifuge briefly. Completely thaw the tube of 10× buffer (provided with the enzyme), and vortex to mix. Do not use buffers that contain ice particles, as the buffer and salt concentrations will not be correct. Briefly centrifuge the tubes of buffer and enzyme; store on ice. (Refer to the section on Proper Enzyme Usage in Appendix II.)

☐   3. Label a 1.5-ml microcentrifuge tube (e.g., chromosomal DNA + *Hin*d III); include the date and your group number or initials. Based on your calculations in step 1, add the reagents to this tube in the order listed. Use a new tip for each reagent. The enzyme is provided in a glycerol storage solution; pipette it slowly and carefully.

☐   4. Prepare a tube of undigested control DNA that contains the same amount of chromosomal DNA and buffer, but no *Hin*d III enzyme. Adjust the volume of dH$_2$O to provide a total volume of 20 μl.

☐   5. Flick each tube to mix, and spin briefly. Incubate both tubes at 37°C for at least 4 hours. If the agarose gel will not be run in the same lab session, store the tubes in the freezer until the next lab period.

VERIFY CLEAVAGE OF DNA BY GEL ELECTROPHORESIS

☐   1. Cast a 0.7% agarose gel, as described in Appendix II. (You may want to run the samples from Lab 10A and Lab 10B at the same time.)

☐   2. Thaw your samples of *Hin*d III-digested and undigested chromosomal DNA, if necessary. Vortex to mix, and centrifuge briefly.

☐   3. Label two 1.5-ml tubes and transfer 5 μl of each of the samples to the appropriate tube. Add 5 μl of 2× DNA Gel Loading Buffer to each tube. Vortex to mix, and centrifuge briefly. Save the rest of your *Hin*d III-cleaved chromosomal DNA for Lab 10C.

☐   4. Load the entire volume of each sample into adjacent wells of the gel. Load DNA molecular size markers (e.g., 1 kb DNA Ladder) in another well of the gel. Keep a record of how the samples were loaded on the gel.

☐   5. Run the gel at 80–90 volts until the blue dye has migrated about halfway down the gel. *WEAR GLOVES WHEN HANDLING SOLUTIONS OF ETHIDIUM BROMIDE, AND WEAR EYE PROTECTION WHEN USING UV LIGHT.* Stain the gel with ethidium bromide as described in Appendix II. Photograph the gel; label and place the photograph in your notebook.

DATA ANALYSIS

- Determine whether your chromosomal DNA was cleaved into fragments with *Hin*d III. Does the pattern of fragments seen on this gel resemble that seen following cleavage with *Hin*d III in Lab 9A? The uncleaved control chromosomal DNA will migrate very slowly and should appear as a distinct band near the top of the gel (refer to Lab 7).

LAB 10B

Cleavage of Plasmid DNA

BACKGROUND

Plasmids, which are circular molecules of DNA that replicate independently of the bacterial chromosome, are often used as vectors for cloning fragments of foreign DNA. Replication of plasmid DNA is carried out by the same bacterial enzymes used to duplicate the bacterial chromosome. DNA replication originates at a specific sequence in the plasmid DNA called the *origin of replication.* Many plasmids used for cloning can replicate many times; some reach copy numbers as high as 700 per cell.

Plasmid cloning vectors have been engineered to contain a region of closely arranged restriction enzyme cleavage sites for easy insertion of foreign DNA. This region is called the *multiple cloning region* (MCR). This region contains multiple restriction enzyme cleavage sites (e.g., 6–12) so that a researcher can choose among a variety of restriction enzymes for cloning. Additionally, the cleavage sites in the MCR are unique, i.e., not present elsewhere in the plasmid.

Plasmid cloning vectors generally contain at least one antibiotic resistance gene that can be used to select bacterial cells that have taken up the plasmid. Antibiotic resistance genes code for proteins that either inactivate the antibiotic or prevent entry of the antibiotic into the bacterial cell. Thus, cells that have been transformed with a plasmid containing such a selectable marker can be grown in the presence of the antibiotic, whereas those without the plasmid will be killed by the antibiotic.

It is desirable to be able to identify colonies that contain recombinant plasmids (i.e., those with foreign DNA inserted in the MCR). A commonly used method is blue/white color screening. Plasmid vectors have been engineered to contain the part of the *lacZ* gene that codes for the amino-terminal fragment of the enzyme β-galactosidase. This fragment of β-galactosidase, whose synthesis can be induced by a small molecule, will combine with the rest of the protein, produced by the host cell, to form an active enzyme. Beta-galactosidase can be detected by its conversion of a colorless substrate into a blue product. Colonies of cells carrying non-recombinant plasmids have both fragments of β-galactosidase and can convert the colorless substrate into a blue compound, turning the colonies blue. Insertion of foreign DNA into the MCR disrupts the reading frame of the *lacZ* gene, and no amino-terminal fragment of β-galactosidase is made. Since no active enzyme can be formed, cells carrying recombinant plasmids cannot convert the substrate to the blue compound and they remain white.

Another method to identify a colony with an insert in the MCR is called *positive selection.* This method requires a special plasmid that normally kills the cells. One example of this is the plasmid you will use in this course, pRL498. That plasmid is described in more detail in the next section. Positive selection vectors cannot replicate in most strains of *E. coli* unless there is a fragment of DNA inserted into the MCR. The advantage of positive selection over blue/white screening is that only colonies with an insert in the MCR of the vector can survive. In techniques such as blue/white screening, colonies without an insert in the vector also survive and must be distinguished by their lack of color. This is not difficult when you are inserting a single known fragment of DNA in the vector and need only a single colony with a recombinant plasmid. However, when you are screening a library with thousands of different potential inserts you need thousands of colonies with inserts in order to find the gene of interest. Identifying the colony with the gene of interest using blue/white screening can be very difficult since typically 70–90% of the colonies on the plate do not have an insert in the vector. With a positive selection vector every colony has an insert, although very few have the gene of interest.

LABORATORY OVERVIEW

In this lab, you will prepare your cloning vector by cleaving it with *Hin*d III. Your cloning vector will be the plasmid pRL498, a high copy number plasmid that has been engineered to make insertion of DNA in the MCR essential for survival of the plasmid in *E. coli* (Elhai and Wolk 1988). This characteristic, which confers positive selection for plasmids with inserted DNA, makes this a positive selection vector. The mechanism for positive selection in pRL498 is that plasmids with long segments of palindromic DNA (inverted repeats of DNA) are not viable in most strains of *E. coli*. The palindromic sequence disrupts replication of the plasmid, and as a consequence, most daughter cells do not receive plasmids and are killed by the antibiotic.

A map of pRL498 is shown in Figure 10B.1. The MCR of pRL498 lies between two nearly identical 535 bp segments of DNA, inverted relative to each

FIGURE 10B.1

Map of cloning vector pRL498. The positive selection vector pRL498 contains several important features. The large inverted repeat regions ensure that the plasmid cannot replicate in most strains of *E. coli*. The multiple cloning region (MCR) provides restriction sites for insertion of foreign DNA, which interrupts the inverted repeats, allowing the plasmid to replicate. The kanamycin resistance gene provides selection for transformants that contain the plasmid. The plasmid can replicate using the origin of replication called *oriV*.

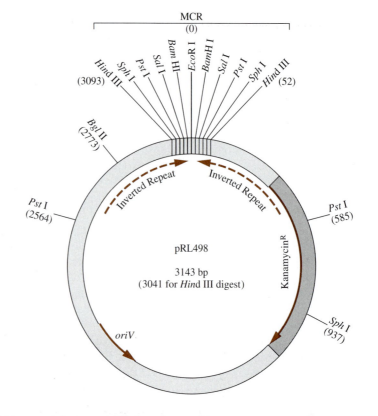

other. The long palindromic segments of DNA disrupt replication of the plasmid in most strains of *E. coli*. (This plasmid can be propagated in strains of *E. coli* having a defect in recombination, however). If a fragment of DNA is inserted in the MCR of pRL498, then the long palindrome is interrupted, and the recombinant plasmid can replicate in *E. coli*. Since pRL498 contains the kanamycin resistance gene (Km^R), the only cells that survive in the presence of kanamycin are those that contain a recombinant plasmid.

(*Note:* The *amyE* gene has been cloned into several more commonly used plasmid cloning vectors, but the plasmids containing the *amyE* gene are extremely unstable and difficult to maintain. It appears that the success of this project requires the use of a positive selection vector.)

SAFETY GUIDELINES

Boiling agarose can cause burns. Wear protective gloves when handling hot agarose solutions.

The electric current in a gel electrophoresis device is extremely dangerous. Never remove the lid or touch the buffer once the power is turned on. Make sure that the counter where the gel is being run is dry.

Ethidium bromide is a strong mutagen and a possible carcinogen. Gloves should always be worn when handling gels or buffers containing this chemical.

UV light, used to illuminate the DNA dyed with ethidium bromide, is dangerous. Protect your eyes and face by wearing a UV-blocking face shield.

PROCEDURE

CLEAVE PLASMID DNA WITH *HIND* III

☐ 1. Determine the amount of each reagent to add to your reaction tube; the final volume of the reaction will be 20 μl. Calculate the volume of pRL498 plasmid DNA that contains 0.5 μg of DNA (your instructor will provide you with the concentration of the plasmid). Calculate the volume of dH$_2$O needed to bring the total volume of the reaction to 20 μl. The volumes of buffer and enzyme to be used are listed as follows. Fill in the blanks and record this in your notebook.

dH$_2$O	__ μl
pRL498 plasmid DNA	__ μl (containing 0.5 μg of DNA)
10× *Hin*d III buffer	2 μl (1× final concentration)
*Hin*d III	1 μl
Total volume	20 μl

☐ 2. Completely thaw the tube of pRL498 DNA (provided by your instructor), vortex to mix, and centrifuge briefly. Completely thaw the tube of 10× buffer (provided with the enzyme), vortex to mix, centrifuge briefly, and store on ice. Do not use buffers that contain ice particles, as the buffer concentration will not be correct. Briefly centrifuge the tube of *Hin*d III enzyme, and keep on ice. (Refer to the section on Proper Enzyme Usage in Appendix II.)

☐ 3. Label a 1.5-ml microcentrifuge tube (e.g., pRL498 + *Hin*d III); include the date and your group number or initials. Based on your calculations in step 1, add the reagents to this tube in the order listed. Use a new tip for each reagent. The enzyme is provided in a glycerol storage solution; pipette it slowly and carefully.

☐   4. Prepare a tube of undigested control DNA that contains the same amount of plasmid DNA and buffer, but no *Hin*d III enzyme. Adjust the volume of dH$_2$O to provide a total volume of 20 μl.

☐   5. Flick each tube to mix, and spin briefly. Incubate both tubes at 37°C for at least 4 hours. If the agarose gel will not be run in the same lab session, store the tubes in the freezer until the next lab period.

VERIFY CLEAVAGE OF DNA BY GEL ELECTROPHORESIS

☐   1. Cast a 0.7% agarose gel, as described in Appendix II. (You may want to run the samples from Lab 10A and Lab 10B at the same time.)

☐   2. Thaw your samples of *Hin*d III-digested and undigested pRL498 plasmid DNA, if necessary. Vortex to mix, and centrifuge briefly.

☐   3. Label two 1.5-ml tubes and transfer 5 μl of each of the samples to the appropriate tube. Add 5 μl of 2× DNA Gel Loading Buffer to each tube. Vortex to mix, and centrifuge briefly. Save the rest of your *Hin*d III-cleaved pRL498 plasmid DNA for Lab 10C.

☐   4. Load the entire volume of each sample into adjacent wells of the gel. Load DNA molecular size markers (e.g., 1 kb DNA Ladder) in another well of the gel. Keep a record of how the gel was loaded.

☐   5. Run the gel at 80–90 volts until the blue dye has migrated about halfway down the gel. ***WEAR GLOVES WHEN HANDLING SOLUTIONS OF ETHIDIUM BROMIDE, AND WEAR EYE PROTECTION WHEN USING UV LIGHT.*** Stain the gel with ethidium bromide as described in Appendix II. Photograph the gel; label the photo and place it in your notebook.

DATA ANALYSIS

- Determine whether the plasmid DNA was linearized with *Hin*d III by comparing the lanes with cut and uncut DNA. The uncut control plasmid DNA may run as several bands on the gel, with the different bands representing supercoiled, relaxed, or nicked circular DNA. Since the mobility of circular DNA differs from that of linear DNA of the same size, none of these bands will appear to be of the correct size (3.1 kb). The cleaved plasmid DNA should run as a single band of about 3 kb since it is a linear piece of DNA. The 100 bp *Hin*d III fragment, released from the MCR, will be too faint to be seen on the gel. None of the bands visible in the lane with the control DNA should be present in the lane with the *Hin*d III-cleaved plasmid DNA. What can you conclude if some of these bands are present in the lane with the cleaved plasmid DNA?

LAB 10C

Ligation of Chromosomal and Plasmid DNA

BACKGROUND

One of the most important steps in DNA cloning is the joining of DNA from different sources to create recombinant molecules. DNA ligase is an enzyme that joins ends of DNA by forming a phosphodiester bond between the $5'$-phosphate of one end of DNA and the $3'$-hydroxyl group of the second end of DNA (Fig. 10C.1). DNA ligase essentially reverses the reaction catalyzed by restriction endonucleases. Any DNA molecules that have been cleaved with restriction enzymes that yield compatible single-stranded ("sticky") ends can be ligated together. The complementary single-stranded ends will hydrogen bond, keeping the two ends of DNA close together long enough for the DNA ligase to form a phosphodiester bond between them. Fragments of DNA with blunt ends can also be ligated; this reaction is much less efficient and requires much higher concentrations of DNA ligase. To function, DNA ligase requires an energy source (i.e., ATP), Mg^{2+}, and optimal buffer conditions. The DNA ligase used most frequently in gene cloning experiments is T4 DNA ligase, which is from the bacteriophage T4.

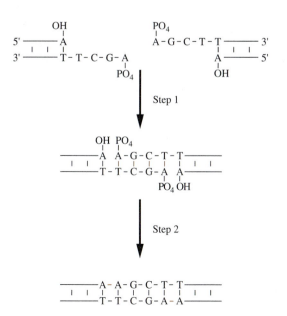

FIGURE 10C.1

Ligation of two strands of DNA cleaved with *Hin*d III. Step 1: hydrogen bonding of the overhanging bases of the complementary ends. Step 2: formation of the phosphodiester bonds.

When ligating fragments of DNA into a cloning vector, the goal is to create recombinant molecules in which each vector contains one insert of DNA. When the fragments of foreign DNA and the vector all have the same compatible ends, several products can be formed. In addition to the desired product (one insert per vector), more than one fragment could be inserted into the vector, the vector could self-ligate, several vectors could join together, etc. The success of a ligation reaction depends upon the relative concentrations of insert and vector DNA. If the total concentration of DNA is low, the two ends of DNA that join will most likely be on the same fragment. At higher concentrations of DNA, there is an increased probability that the two ends that join will be from different molecules of DNA, leading to the formation of recombinant molecules. If the ratio of insert to vector DNA increases further, there is an increased likelihood that multiple fragments will be inserted into a vector. For most ligation reactions, a onefold to threefold molar excess of insert to vector DNA is ideal.

LABORATORY OVERVIEW

A ligation reaction will be set up with the *Hin*d III restriction fragments of *B. licheniformis* chromosomal DNA (Lab 10A) and the *Hin*d III-digested plasmid pRL498 (Lab 10B). Many recombinant plasmids will be created. A few of these will contain the 3.3 kb *amyE* gene fragment, but most will contain other fragments of *B. licheniformis* DNA. Most of these recombinant plasmids should have one fragment of DNA inserted into the MCR, but some may have more than one inserted fragment. In fact, about half of the *amyE*$^+$ plasmids tend to also have another insert. It is possible that recombinant pRL498 plasmids with large segments of inserted DNA interrupting the palindromic DNA have an advantage over those with smaller inserts. To decrease the number of the *amyE*$^+$ plasmids with extra inserted DNA, the ratio of insert DNA to vector DNA will be about one to one during the ligation reaction. It is possible that a lower ratio of insert to vector (i.e., one to two) would decrease the number of *amyE*$^+$ plasmids with extra inserted DNA, but it would also reduce the number of *amyE*$^+$ plasmids (which average around 0.1% of the total recombinant clones).

PROCEDURE

1. Thaw your samples of *Hin*d III-cleaved chromosomal DNA (Lab 10A) and *Hin*d III-cleaved pRL498 (Lab 10B), if necessary. Incubate each at 65°C for 20 minutes to inactivate the restriction enzyme so it cannot recleave the DNA that is joined during the ligation reaction. Centrifuge each briefly to collect the entire sample in the bottom of the tube.

2. Calculate the volume of digested chromosomal DNA that contains 100 ng of DNA. In Lab 10A, you added 0.5 μg (500 ng) of chromosomal DNA into a final volume of 20 μl. Thus, the concentration of the *Hin*d III-cleaved chromosomal DNA is 500 ng/20 μl = 25ng/μl. Refer to the formulas presented on p. 101, if necessary, to determine the volume that contains 100 ng of DNA.

   Calculate the volume of digested plasmid DNA that contains 100 ng of DNA, as described previously.

   Add your values to the list that follows and record this in your notebook. Most buffers supplied with T4 DNA ligase contain ATP, but since ATP is susceptible to degradation after repeated cycles of freezing and thawing, you will supplement the reaction with ATP.

Digested chromosomal DNA	____ μl
Digested plasmid DNA	____ μl
10$\times$ ligase buffer	1 μl
20 mM ATP	0.5 μl
T4 DNA ligase	0.5 μl
Total volume	10 μl

3. Completely thaw the tube of 10$\times$ ligase buffer. Do not use buffers that contain ice particles, as the buffer and salt concentrations will not be correct. Vortex to mix, centrifuge briefly, and store on ice. Briefly centrifuge the tubes of T4 DNA ligase and ATP, and store on ice. (Refer to the section on Proper Enzyme Usage in Appendix II.)

4. Label a 0.5-ml microcentrifuge tube (e.g., ligation); include the date and your group number or initials. Following the list you completed in step 2, add the reagents to this tube in the order listed. Use a new tip for each reagent. The enzyme is provided in a glycerol storage solution; pipette it slowly and carefully. Flick the tube to mix, and centrifuge briefly.

5. Incubate the reaction at 16°C for at least 8 hours (this can be done in a thermocycler). The ligation reaction should be stored in the freezer until needed for the transformation protocol of Lab 10D.

DATA ANALYSIS

- You will be able to determine whether your ligation reaction was a success after the transformation protocol of Lab 10D. Since all of the DNA fragments in the ligation reaction have identical sticky ends, several possible reactions can occur: (1) plasmid molecules can self-ligate without an insert; (2) chromosomal DNA fragments can self-ligate, or can ligate with each other; and (3) a plasmid molecule can ligate with one or more chromosomal fragments. Based upon what you know about pRL498, as well as plasmids in general, predict what would happen to these three types of molecules if they were introduced into *E. coli*.

LAB 10D

Transformation

BACKGROUND

To complete the process of DNA cloning, newly formed recombinant plasmid DNA molecules must be introduced into bacterial cells for replication. This results in the production of many identical copies, or clones, of the recombinant plasmid molecule. *Transformation* is the process whereby DNA is taken inside competent bacterial cells. *E. coli* cells are made competent to take up foreign DNA by treatment with ice-cold solutions of divalent cations or other chemicals that weaken the cell wall. The competent cells are mixed with the foreign DNA and then subjected to a brief pulse of heat, during which time they can take up the exogenous DNA. Only a small percentage of the competent cells take up the DNA and become transformed. Selectable markers encoded by the plasmid are used to identify those transformants. If the cloning vector carries an antibiotic resistance gene, then the transformed cells can be selected by their ability to grow in the presence of the antibiotic. First, however, the cells must be incubated in a rich medium for a short period of time (without the antibiotic) so that the antibiotic resistance gene can be expressed in those cells that have been transformed.

LABORATORY OVERVIEW

In this lab, you will use commercially prepared supercompetent cells so that your library of *B. licheniformis* chromosomal DNA will contain as many recombinant plasmids as possible. Commercially prepared supercompetent cells yield $\geq 10^9$ transformed colonies/μg of plasmid DNA, whereas "homemade" competent cells yield only 10^6–10^7 transformed colonies/μg of plasmid DNA. Since the *amyE* gene represents only a small portion of the total chromosome, many recombinants must be screened to find one containing the gene of interest.

The supercompetent cells will be mixed with some of your ligation reaction from Lab 10C, heat shocked, briefly incubated in a rich medium, and then spread on LB agar plates that contain kanamycin and starch. Since pRL498 is a positive selection vector that contains the Km^R gene, only *E. coli* cells that have been transformed with recombinant pRL498 molecules will be able to survive on the medium containing kanamycin. Most of these transformants will not contain the *amyE* gene, but rather will have other fragments of chromosomal DNA inserted in the MCR of pRL498. Since the frequency of colonies that produce α-amylase averages about 0.1% of all the colonies that grow on the selection plates, you will have to screen more than 1000 colonies to find one that contains the *amyE* gene.

In Lab 10E, you will screen for those that contain the 3.3 kb *amyE* gene fragment by their ability to degrade starch. The plates will be stained with Lugol

Solution (as in Lab 1); the plates will turn purple everywhere except around those colonies with the *amyE* gene. There will be a halo of clearing around these colonies because they released α-amylase, which degraded the starch.

SAFETY GUIDELINES

Use sterile technique when handling culture media and *E. coli* cells. Discard all contaminated tubes and tips into waste bags, which will be autoclaved prior to disposal.

PROCEDURE

The procedure is based on Stratagene's SoloPack™ Gold Supercompetent *E. coli* cells, which are provided in 50 μl single-reaction aliquots. If supercompetent cells are used from another supplier, follow the transformation protocol provided with those cells.

   To ensure that everything is working properly, a positive control transformation should be performed (one for the entire class is sufficient). This control also will allow you to calculate the transformation efficiency of these cells. The steps of the control transformation are identical to those listed in the following steps, except where indicated.

☐ 1. **Review the section on Sterile Technique in Appendix II before starting this lab.** Thaw your ligation mixture from Lab 10C, centrifuge briefly, and store on ice.

☐ 2. Thaw a tube of supercompetent *E. coli* cells on ice (this takes 5 minutes or less). ***KEEP THE CELLS ON ICE AT ALL TIMES.*** Once the cells have thawed, swirl the tube gently to mix the cells.

   The group performing the positive control reaction should thaw another tube of cells.

☐ 3. Add 1 μl of β-mercaptoethanol (provided with the cells) to the tube of cells (control cells too). Swirl the tube gently to mix the cells. Incubate the cells on ice for 10 minutes, swirling the tube every 2 minutes.

☐ 4. Add 2 μl of your ligation mixture to the tube of cells and swirl the tube gently to mix. Incubate on ice for 20 minutes.

   Positive control: Dilute the control plasmid (pUC18) provided with the cells tenfold with sterile dH$_2$O. Add 1 μl (contains 0.01 ng of DNA) of this diluted control plasmid to the control cells. Incubate on ice for 20 minutes.

☐ 5. Heat shock the cells by placing the tube of cells in a water bath set to 54°C for **exactly 60 seconds (i.e., 1 minute).** Remove any ice trapped under the bottom of the tube before placing it in the water bath. Do not shake the tube.

✺ CAUTION: The temperature and duration of the heat pulse are critical for obtaining the highest transformation efficiency.

☐ 6. Place the tube of cells on ice and incubate for 2 minutes.
☐ 7. Add 150 μl of NZY$^+$ broth (a rich medium that contains salts, protein extracts, and glucose) to the tube of cells and incubate at 37°C for 1 hour while shaking at 225–250 rpm.

8. While waiting, label five LB agar plates containing 50 μg/ml of kanamycin and 0.5% starch (L agar + Km + starch plates); include your group number or initials on each plate.

    Positive control: Label one LB agar plate containing 50 μg/ml of ampicillin (the plasmid pUC18 contains the ampicillin resistance gene).

9. Place ~40 μl of the cell mixture on one of your plates. Spread the cell mixture with a sterile cell spreader. If you are using a bent glass rod as a spreader, sterilize it by dipping it into 95% ethanol, draining off the excess ethanol, and passing it through a flame. Allow the glass rod to cool for about 30 seconds before using it to gently but thoroughly spread the cells over the entire surface of the plate. Repeat with the other four plates, using all of your transformation mixture. Sterilize the spreader each time.

    Positive control: First place 200 μl of NZY$^+$ broth on the control LB agar + ampicillin plate, and then place 5 μl of your control transformation in the pool of broth. Use a sterile spreader to spread the mixture over the entire plate.

10. Incubate the plates (inverted) at 37°C until the colonies are at least 1 mm in diameter. This usually takes 20–24 hours.

    The plates can then be stored at 4°C, but only for about 1 day. When stored longer, the zones of clearing become too large to easily identify which clones produced α-amylase.

DATA ANALYSIS

- Each experimental plate should have 300–1000 transformants. If you have transformants, what can you conclude about the success of the ligation reaction (Lab 10C)?

- Estimate the total number of transformants on your plates. If all of your plates have about the same density of colonies, you can count those on one plate and use that number to estimate your total number of colonies. To do this, use a marking pen to divide one plate into four equal sections and then count the colonies in one section. Multiply that number by 4 to estimate the total number of colonies on one plate. Multiply that number by the total number of plates to estimate your total number of transformants. If your plates have drastically different cell densities, then you should count each plate (divide the plate into four sections, etc.).

- The group that performed the positive control transformation should count the number of colonies on the positive control plate. Share this number with the rest of the class so that all can determine the efficiency of transformation of the supercompetent cells using the following formula:

$$\frac{\text{number of colonies}}{\text{ng of plasmid used in transformation}} \times \frac{1000 \text{ ng}}{\mu\text{g}} \times \text{dilution factor}$$

$$= \frac{\text{transformed colonies}}{\mu\text{g}}$$

    The amount of control plasmid used in the transformation was 0.01 ng. Since 5 μl of the 200 μl transformation reaction was plated, the dilution factor is 200/5 = 40.

- Compare the efficiency of transformation of the control plasmid with the efficiency claimed by the manufacturer of the supercompetent cells (generally $\geq 10^9$ transformants/μg of DNA). Did you get the efficiency predicted by the manufacturer?

LAB 10E

Identification of α-Amylase Clones

LABORATORY OVERVIEW

The colonies of transformed cells, or clones, on your plates represent a library of *B. licheniformis* chromosomal DNA. Only a few (~0.1%) of the clones in this library will contain the 3.3 kb *Hind* III fragment that contains the *amyE* gene. The rest of the clones (~99.9%) will contain other *Hind* III fragments of *B. licheniformis* DNA. In order to identify those clones that have the *amyE* gene, you will screen all of your transformants for those few that show α-amylase activity as determined by their ability to degrade starch.

Production of α-amylase in *E. coli* assumes that the 3.3 kb *Hind* III fragment of *B. licheniformis* DNA contains the coding region as well as the control regions of the *amyE* gene. To get expression, the RNA polymerase of *E. coli* has to be able to recognize the promoter of the *amyE* gene to initiate transcription. Additionally, the ribosomes of *E. coli* have to recognize the sequence in the mRNA just upstream of the start codon (ribosome binding site) to initiate translation. The end result is production of the *B. licheniformis* α-amylase enzyme in *E. coli*. Surprisingly, this α-amylase protein is secreted by *E. coli* cells (which do not normally secrete proteins) by a mechanism that is not understood.

In Part I of this lab, you will screen your entire library of transformants for colonies that produce α-amylase by their ability to degrade starch. Lugol iodine staining will reveal zones of clearing around any transformants that have α-amylase activity, thus identifying clones that are likely to have the *amyE* gene (Fig. 10E.1). This is called a *primary screening*. Because there may be several colonies close together in a zone of clearing on the original transformation plates, you will do a *secondary screening* (Part II) to make sure the correct colony is isolated. Putative *amyE*⁺ colonies will be picked from the original plates; these will be diluted and plated on fresh LB agar + Km + starch plates. In Part II of this lab, these plates will be screened for colonies that produce α-amylase, as revealed by zones of clearing following Lugol iodine staining. One or two well isolated, *amyE*⁺ colonies will be selected from the secondary screening plates for further analysis in Lab 11.

SAFETY GUIDELINES

Use sterile technique when handling culture media and *E. coli* cells. Discard all contaminated tubes and tips into waste bags, which will be autoclaved prior to disposal.

FIGURE 10E.1

Iodine-stained starch plate with transformants. Arrow points to one colony that displays clearing of starch, indicating the likely presence of the recombinant *amyE*⁺ plasmid.

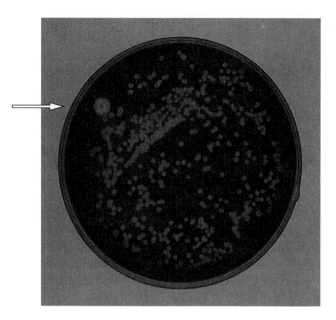

Part I: Primary Screening

PROCEDURE

1. ***USE GOOD STERILE TECHNIQUE.*** Working with one plate at a time, gently flood the surface of each plate with 2–3 ml of Lugol Solution and wait until the agar turns purple. ***DO NOT SWIRL THE PLATES AS THIS WILL DISLODGE THE CELLS.*** Gently pour off and discard the Lugol Solution and carefully inspect each plate for colonies that have a halo of clearing. It helps to view the plates against a white background. Circle putative *amyE*⁺ colonies on the bottom of the plate with a marking pen. Record in your notebook the number of *amyE*⁺ colonies on each plate.

   Note: You can restain the plates with Lugol Solution and photograph them after you pick the putative *amyE*⁺ colonies. It is not necessary to photograph every plate; choose the best-looking one. Label the photo and place it in your notebook. (When you are finished with your original plates, wrap them in Parafilm and store them at 4°C until Lab 10E is completely finished.)

2. Allow the surface of the plate to dry for several minutes, and then use a sterile pipette tip or inoculating loop to pick one or two putative *amyE*⁺ colonies. Transfer each colony to a 1.5-ml tube containing 1 ml of LB broth + Km. The *amyE*⁺ colony should lie in the center of the halo of clearing. If there are several colonies close together in the center of the clear zone and it is difficult to determine which could be the *amyE*⁺ colony, pick all of them since your secondary screening will separate the *amyE*⁺ colony from the other colonies. Label each tube of cells with a name or number. Vortex each tube to mix the cells and broth.

   If you have only one putative *amyE*⁺ colony and another group has more than two putative *amyE*⁺ colonies, you may wish to pick one of their extra ones to ensure that you will have a clone for future experiments.

FIGURE 10E.2
Serial dilutions and plating.

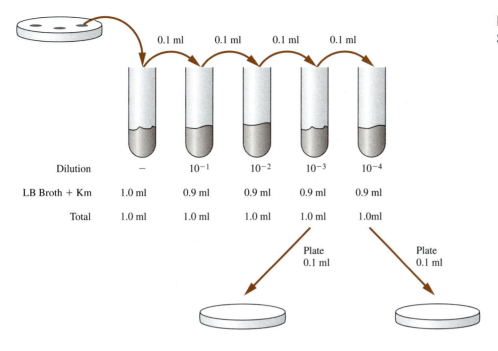

3. For each colony you picked, serially dilute the cells as diagrammed in Figure 10E.2. Label four sterile 1.5-ml tubes as 10^{-1}, 10^{-2}, 10^{-3}, and 10^{-4}. Add 0.9 ml of LB broth + Km to each tube. Add 0.1 ml of the cells from the original tube (from step 2) to the tube labeled "10^{-1}." Vortex to mix. Add 0.1 ml of the 10^{-1} diluted cells to the tube labeled "10^{-2}." Vortex to mix. Continue diluting the cells in this manner.

4. For each colony you picked, you will plate cells from the 10^{-3} and 10^{-4} dilutions (see Fig. 10E.2) to optimize your chances of getting well-isolated $amyE^+$ colonies. If one dilution results in too many (or too few) colonies, the other dilution should give a reasonable number. Place 100 μl of the 10^{-3} dilution on a new LB agar + Km + starch plate. Use a sterile cell spreader to spread the mixture over the surface of the plate, as described in Lab 10D. Place 100 μl of the 10^{-4} dilution onto another LB agar + Km + starch plate and spread as previously described. Label each of these secondary plates with the name of the original colony and the dilution factor.

5. Incubate the secondary plates (inverted) at 37°C until the colonies are 1–2 mm in diameter, usually about 18–24 hours.
   Note: If the plates will not be screened within the next day or two, store the freshly inoculated plates at 4°C until the day before the next lab period, at which time they should be placed at 37°C. This prevents the zones of clearing from becoming too large.

DATA ANALYSIS

- From your original transformation plates, determine your total number of $amyE^+$ colonies. (Present your data in a table, showing the number of $amyE^+$ colonies per plate.) If you have more than one positive $amyE^+$ colony, do they all display zones of clearing of the same size? Are some zones of clearing larger? Using your estimate of the total number of transformants from Lab 10D, determine the percentage of $amyE^+$ colonies in your library of *B. licheniformis* DNA.

Part II: Secondary Screening

PROCEDURE

You will screen the secondary plates and pick a well-isolated $amyE^+$ colony corresponding to each of your original isolates. For each of the clones you pick from your secondary screening plates, you will streak some of these cells onto LB agar + Km plates so that a constant supply of cells will be available for future experiments. (These cells should be restreaked onto fresh plates every 2 weeks to maintain working stock plates.) You will also use some of these $amyE^+$ cells for the PCR reactions in Lab 11A.

1. *USE GOOD STERILE TECHNIQUE.* Label a 1.5-ml sterile microcentrifuge tube for each of the original colonies you picked (Part I, preceding section). Add 200 μl of sterile dH_2O to each tube.

2. Gently flood the surface of each secondary plate with 2–3 ml of Lugol Solution and wait until the agar turns purple. *DO NOT SWIRL THE PLATES AS THIS WILL DISLODGE THE CELLS.* Gently pour off and discard the Lugol Solution and carefully inspect the plate for colonies that have a halo of clearing. Circle the positive $amyE^+$ colonies on the back of each plate with a marking pen. These plates should have many $amyE^+$ colonies. In fact, all of the colonies could be $amyE^+$ if the original colony you picked was very well isolated. Most of the plates will contain a mixture of $amyE^+$ colonies and negative colonies. It is also possible that none of the colonies will be positive if you picked the wrong colony from the original transformation plate.

3. Allow the surface of the agar to dry, and for each of the original colonies you picked, use a sterile pipette tip or inoculating loop to pick one well-isolated $amyE^+$ colony from either the 10^{-3} or 10^{-4} dilution plate. Transfer the cells to the appropriately labeled microcentrifuge tube. Vortex to mix.

4. Prepare a stock plate for each of your putative $amyE^+$ clones. Label and date an LB agar + Km plate (no starch) for each. Dip a sterile inoculating loop into the tube of cells and quadrant streak the cells (see pp. 10–11) onto the appropriate plate to produce single colonies. Incubate the plate(s) overnight at 37°C, and then store at 4°C. Store your tube of $amyE^+$ cells at 4°C until needed for Lab 11A.

5. You also need some negative cells for Lab 11A and Lab 12. Pick a colony that did not have a halo of clearing from one of your secondary screening plates or your original transformation plates. Transfer the cells to an appropriately labeled 1.5-ml tube that contains 200 μl of sterile dH_2O. Vortex to mix. Quadrant streak some of the negative cells onto a labeled and dated LB agar + Km plate, incubate overnight at 37°C, and then store at 4°C. Store the tube of negative cells at 4°C until needed for Lab 11A.

DATA ANALYSIS

- Summarize your data. Were all of the colonies on your secondary screening plates positive? Did your secondary plates contain a mixture of positive and negative cells? Were you able to isolate a colony of $amyE^+$ cells for each of your original picks from the transformation plates (Part I)? What can you conclude about your primary pick if your secondary screening plates contained only negative cells?

QUESTIONS

1. Explain why plasmid vectors used for DNA cloning need an origin of replication, a multiple cloning region, and at least one antibiotic resistance gene.
2. Explain the function of DNA ligase. What reagents are required for DNA ligase to function?
3. What is transformation? Explain how DNA is introduced into *E. coli* cells.
4. Which of your transformants contained a recombinant pRL498 plasmid? Explain.
5. Explain the screening procedure used to identify which colonies contained a functional *B. licheniformis amyE* gene.
6. Most of the colonies on your original transformation plates did not digest starch. Describe what foreign DNA (if any) is present in these cells.
7. Write a paragraph that describes the process of gene cloning.

LAB 11

Verification and Mapping of α-Amylase Clones

GOAL

The goal of this lab is to verify that the putative *amyE*$^+$ colonies do indeed contain the *amyE* gene from *B. licheniformis*.

OBJECTIVES

Upon completing Lab 11, you will be able to explain and perform the following techniques:

1. PCR amplification of a cloned gene from transformed cells
2. preparation of *E. coli* cells for permanent stock
3. isolation of plasmid DNA
4. restriction enzyme digestion of plasmid DNA
5. gel electrophoresis of restriction enzyme digested plasmid DNA
6. Southern blot analysis of plasmid
7. mapping a new plasmid

INTRODUCTION

Once a putative recombinant clone has been identified, additional methods are needed to verify that it contains the DNA of interest. The recombinant clone can be screened by PCR or by Southern hybridization for the DNA of interest. Determination of the size of the recombinant plasmid often reveals whether it contains the DNA of interest. Upon confirmation that the recombinant plasmid contains the DNA of interest, the DNA can be analyzed further by mapping the relative location of restriction enzyme cleavage sites within the DNA sequence. In this lab, you will determine whether the putative *amyE*$^+$ colonies you isolated in Lab 10E contain the 3.3 kb *amyE* gene fragment. You will then isolate plasmid DNA from an *amyE*$^+$ clone and map the locations of some restriction enzyme cleavage sites and the location of the *amyE* gene within the 3.3 kb fragment.

GENERAL LABORATORY OVERVIEW

This lab is divided into five sections:

Lab 11A: Putative $amyE^+$ clones will be verified by PCR.

Lab 11B: Plasmid DNA will be isolated from $amyE^+$ clones.

Lab 11C: The size of the $amyE^+$ plasmid will be determined, and the 3.3 kb insert will be mapped by restriction enzyme cleavage.

Lab 11D: A permanent stock of cells containing the $amyE^+$ plasmid will be prepared.

Lab 11E: The location of the $amyE^+$ gene within the 3.3 kb insert will be determined by Southern hybridization.

TIMELINE

Day 1	Lab 10E	Secondary screening
	Lab 11A	Set up PCR reactions (~0.5 hour)
Day 2	Lab 11A	Run PCR samples on a gel (~2 hours)
	Lab 11B	Set up cultures for isolation of plasmid DNA (<0.5 hour)
Day 3	Lab 11B	Isolate plasmid DNA (~1 hour)
	Lab 11C	Set up *Bgl* II cleavage reaction to determine size and orientation of insert (<0.5 hour)
Day 4	Lab 11C	Run *Bgl* II-cleaved DNA on gel (2–2.5 hours)
		Set up restriction enzyme cleavages for mapping (~0.5 hour)
	Lab 11D	Set up cultures for permanent stock (<0.5 hour)
Day 5	Lab 11C	Run fragments of cleaved DNA on gel (2–2.5 hours)
	Lab 11D	Prepare permanent stock (<0.5 hour)
	Lab 11E	Denature DNA and set up Southern transfer (1–1.5 hours)
Day 6	Lab 11E	Affix DNA to membrane and set up Southern hybridization (1.5–2 hours)
	Lab 11C	Construct plasmid map
Day 7	Lab 11E	Detect hybridized DNA (2.5–3.5 hours)
		Finish map of 3.3 kb insert

LAB 11A

Verification of α-Amylase Clones Using PCR

LABORATORY OVERVIEW

PCR will be used to verify that the positive clones isolated in Lab 10E contain the *amyE* gene. The template DNA for PCR will come directly from the cells picked during the secondary screening. A small amount of this cell suspension will be added to the PCR reaction and, upon heating in the thermocycler, the cells will lyse and release the DNA for the reaction. The PCR primers used in Lab 8 will be used for this PCR reaction. Only those cells containing *amyE*$^+$ plasmids will yield the 433 bp PCR product.

SAFETY GUIDELINES

Boiling agarose can cause burns. Wear protective gloves when handling hot agarose solutions.

The electric current in a gel electrophoresis device is extremely dangerous. Never remove the lid or touch the buffer once the power is turned on. Make sure that the counter where the gel is being run is dry.

Ethidium bromide is a strong mutagen. Gloves should always be worn when handling gels or buffers containing this chemical.

UV light, used to illuminate the DNA dyed with ethidium bromide, is dangerous. Protect your eyes and skin by wearing a UV-blocking face shield.

PROCEDURE

ASSEMBLE PCR REACTION TUBES

You will set up a reaction for each of the positive *amyE*$^+$ clone(s) you isolated in Part II of Lab 10E, as well as a reaction using the negative cells you picked in that lab. You will also set up a positive control reaction using the *B. licheniformis* chromosomal DNA that you diluted 400-fold for PCR in Lab 8, and you will set up a negative control reaction without DNA.

- [ ] 1. First you will make your own PCR Master Mix. This mix contains everything needed for PCR (except the DNA) that you will add to each PCR reaction tube in one pipetting step. Table 11A.1 lists the components of the PCR Master Mix, the concentration of the stock solutions, and the volume of each needed for one PCR reaction. In

TABLE 11A.1
PCR Master Mix

Reagent	Concentration	Volume/ Reaction	Conversion Factor	Volume for Master Mix
Sterile dH₂O*		8.0 μl		
PCR buffer*	10×	2.0 μl		
MgCl₂*	25 mM	2.0 μl		
dNTP mix	5 mM	0.8 μl		
Forward Primer	5 μM	1.0 μl		
Reverse Primer	5 μM	1.0 μl		
Taq DNA polymerase	5 U/μl	0.2 μl		
Final volume		15 μl		

*Some PCR buffers already contain MgCl₂. Check with your instructor about the one you are using and adjust the volumes of dH₂O and MgCl₂, if necessary.

principle, you simply multiply the volume/reaction by the number of reactions you are performing to determine how much of each reagent to add to your PCR Master Mix. In reality, you want to make a little extra PCR Master Mix (e.g., another half reaction). Thus, if you are performing 5 PCR reactions, make enough PCR Master Mix for 5.5 PCR reactions. Complete Table 11A.1 to determine how much of each reagent to add to your PCR Master Mix. Multiply the volume/reaction by the conversion factor (# reactions + 0.5 reaction) to calculate the volume of each to add to your PCR Master Mix. Include this table in your notebook.

2. Label a sterile 1.5-ml tube as "PCR Master Mix," and store on ice. Make sure all the reagents are completely thawed. Briefly centrifuge each tube before opening. Keep all of the reagents on ice. Based upon your calculations in step 1, add the reagents to your PCR Master Mix tube in the order listed. Use a new tip for each reagent. Gently flick the tube to mix, and centrifuge briefly. Keep on ice.

3. Label a sterile 0.5-ml tube for each of your experimental and control reactions. Assemble your reactions according to Table 11A.2. First add the PCR Master Mix to each tube. Then add 5 μl of your *amyE*⁺ cells to the appropriate experimental tube(s), and add 5 μl of the negative cells to the negative control tube. Add 1 μl of your 400-fold diluted *B. licheniformis* chromosomal DNA and 4 μl of sterile dH₂O to the positive control tube. Add dH₂O to the no-DNA control tube. Flick each tube to mix, centrifuge briefly, and store on ice.

TABLE 11A.2
PCR Samples

Condition	Volume of Master Mix	Volume of Cells	Volume of dH₂O	Volume of DNA	Final Volume
amyE⁺ experimental	15 μl	5 μl	0	0	20 μl
Negative control	15 μl	5 μl	0	0	20 μl
Positive control	15 μl	0	4 μl	1 μl	20 μl
No-DNA control	15 μl	0	5 μl	0	20 μl

4. If your thermocycler does not have a heated lid, overlay each sample with ~30 μl of sterile mineral oil. Centrifuge briefly. Keep on ice until all the samples are ready to be loaded into the thermocycler.

PERFORM PCR

1. Follow the manufacturer's directions for programming the thermocycler. The steps of this PCR program include the following:

Initial Denaturation:	95°C, 5 min
Cycle 30 times:	
Denature	95°C, 30 sec
Anneal	62°C, 1 min
Extend	72°C, 2 min
Final Extension:	72°C, 10 min
Hold:	12°C, 24 hr

2. Place the samples in the thermocycler, securely close the lid, and start the program.
3. Remove the samples from the thermocycler when the program is finished. If gel electrophoresis will not be performed immediately, store the samples in the freezer until the next lab session.

VERIFY PCR PRODUCT BY GEL ELECTROPHORESIS

1. Cast a 1% agarose gel as described in Appendix II.
2. If your PCR reactions were frozen, place the tubes at room temperature and allow the samples to thaw completely. Centrifuge briefly.
3. Add 3 μl of 6× DNA Gel Loading Buffer to each PCR tube, vortex to mix, and centrifuge briefly. If you covered the samples with mineral oil, transfer the blue aqueous phase of each sample to a new, labeled 1.5-ml tube. For this, carefully place the tip of an adjustable micropipette just below the clear layer of oil into the blue sample and withdraw as much of the sample as possible. Do not draw oil into the tip. If oil does get into the tip, gently expel it back into the tube. If the oil and sample in the tube get mixed, centrifuge the tube to separate the layers.
4. Load the entire volume of each sample into a well of the gel. Record in your notebook the order in which you loaded the samples on the gel. Load DNA molecular size markers on the gel.
5. Run the gel at ~90 volts until the blue dye has migrated about halfway down the gel. (The PCR product migrates within the blue dye.) *WEAR GLOVES WHEN HANDLING ETHIDIUM BROMIDE SOLUTIONS, AND WEAR EYE PROTECTION WHEN USING UV LIGHT.* Stain the gel with ethidium bromide as described in Appendix II. Photograph the gel. Label this photo and place it in your lab notebook.

DATA ANALYSIS

- Interpret your data. Which of your PCR samples had a product? Was the product the expected size? Determine which of the positive clones that you picked in Lab 10E contain the *amyE* gene. Discuss the results of the control PCR reactions.

LAB 11B

Isolation of Plasmid DNA from α-Amylase Clones

BACKGROUND

When a new recombinant clone is isolated, one must verify that the correct fragment of DNA was inserted in the vector. Since most cloning vectors are high copy number plasmids, sufficient plasmid can be obtained from a small amount of bacterial culture to perform the initial plasmid characterization. This isolation of plasmid DNA from a small culture of bacterial cells is often called a *miniprep*.

The isolation of plasmid DNA takes advantage of the small size and the closed circular structure of plasmid DNA. The *E. coli* cells are exposed to a solution of SDS and NaOH, which not only causes the cells to lyse but also denatures the proteins and DNA. Upon denaturation, DNA becomes single-stranded but the two strands of a circular piece of DNA remain linked (like two links in a chain). The cell lysate is then neutralized with potassium acetate, which allows the separated strands of DNA to base pair. Since the two strands of the small plasmid DNA are in close proximity, they can readily find their complement, base pair, and form intact plasmid molecules. The long strands of chromosomal DNA cannot easily find their complement, and they get tangled and pulled out of solution with the precipitated proteins during the subsequent centrifugation step. The small plasmid DNA remains in the supernatant and can be purified further by either organic extractions or binding to a special resin.

LABORATORY OVERVIEW

To isolate plasmid DNA, you first have to grow an overnight culture of your recombinant strain. You will isolate plasmid DNA from a small volume (~3 ml) of liquid culture using a commercial kit (i.e., Wizard® *Plus* Minipreps DNA Purification System from Promega, Inc.). The cells will be lysed and the chromosomal DNA will be selectively precipitated as described previously. You will use a DNA purification resin to purify the plasmid DNA from the supernatant. Other contaminants from the supernatant will be washed from the resin with a wash solution, and the plasmid DNA will be eluted from the resin with a small volume of buffer containing EDTA, which chelates the Mg^{++} required by nucleases.

SAFETY GUIDELINES

Use sterile technique when setting up cultures and handling *E. coli* cells. Discard all contaminated tubes and tips into waste bags, which will be autoclaved prior to disposal.

PROCEDURE

INOCULATE CULTURES

You should isolate plasmid DNA from a freshly grown culture of *E. coli.* You can either inoculate and incubate the broth the day before the miniprep lab or store the inoculated broth at 4°C until the day before the miniprep lab, at which time it should be incubated overnight at 37°C.

☐ 1. ***USE GOOD STERILE TECHNIQUE.*** Label a sterile 15-ml culture tube for each clone that was verified by PCR (Lab 11A) to have the *amyE* gene. Use a sterile serological pipette to add 4 ml of LB broth + Km to each tube.

☐ 2. Recover the stock plates on which you streaked your putative *amyE$^+$* colonies in Lab 10E. Use a sterile inoculating loop to transfer one colony from the appropriate stock plate to the corresponding tube.

☐ 3. Incubate the tubes at 37°C overnight while shaking (~225 rpm).

ISOLATE PLASMID DNA

The protocol presented is for the Wizard® *Plus* Minipreps DNA Purification Kit from Promega, Inc. If you are using a kit from a different manufacturer, follow the instructions provided with that kit.

☐ 1. Gently vortex each tube of *E. coli* culture to resuspend cells. Transfer ~1.5 ml of cells to a labeled, 1.5-ml microcentrifuge tube and close tightly. (Use a separate tube for each culture.)

☐ 2. Centrifuge for 1 minute in a balanced microcentrifuge at full speed.

☐ 3. Pour out and discard the broth into a container for liquid biological waste. Do not dislodge or discard the pellet of cells. Fill the same tube with another 1.5 ml of the culture. **If you are processing two separate cultures, be careful not to mix them.**

☐ 4. Centrifuge for 1 minute in a balanced microcentrifuge at full speed.

☐ 5. Pour out and discard the broth into a container for liquid biological waste. Use a pipette to remove any remaining liquid, leaving the cell pellet in the bottom of the tube.

☐ 6. Add 200 μl of Cell Resuspension Solution (50 mM Tris-HCl, pH 7.5, 10 mM EDTA, 100 μg/ml RNase A) to the tube, and vortex vigorously to resuspend the cells. Turn the tube upside down and flick with your finger to spread the cell suspension along the walls of the tube; there should be no sign of the pellet or clumps of cells, only a homogeneous suspension. If there are clumps, vortex again.

☐ 7. Add 200 μl of Cell Lysis Solution (1% SDS, 0.2 M NaOH). Cap the tube and mix by inverting gently four times. **Too much agitation may result in shearing of the chromosomal DNA, which will then co-purify with the plasmid DNA.** Opening the lid should expose a viscous, sticky solution.

☐ 8. Add 200 μl of Neutralization Solution (1.32 M potassium acetate, pH 4.8). Cap and mix by inverting the tube gently four times. A white precipitate should be visible in the tube.

☐ 9. Centrifuge the tube at full speed for 5 minutes. After centrifugation, there should be a tight white pellet at the bottom and along one wall of the tube and a clear supernatant. If a pellet has not formed, centrifuge the tube at full speed for another 15 minutes.

☐ 10. Prepare a Wizard® minicolumn. Remove the plunger from a 3-ml disposable syringe. Attach the syringe barrel to the Luer-Lok extension of the minicolumn. Thoroughly mix the DNA Purification Resin and add 1 ml of the resin to the barrel of the minicolumn/syringe assembly.

☐ 11. Without disturbing the pellet, carefully pour the supernatant from step 9 into the syringe barrel of the minicolumn/syringe assembly. ***DO NOT TAP THE TUBE.*** Alternatively, you can use a micropipette to transfer the supernatant solution. Do not transfer the pellet or any of the precipitated material along the wall of the tube (or floating in the tube) since this could result in contamination of the plasmid DNA. It is better to leave some of the supernatant rather than contaminate your DNA preparation. Discard the tube containing the pellet.

☐ 12. Carefully insert the plunger into the barrel of the minicolumn/syringe assembly and gently push the slurry into the minicolumn. Discard the solution that flows through the column.

☐ 13. Remove the syringe from the minicolumn by twisting them apart. Then remove the plunger from the syringe barrel. It is important that you perform these two manipulations in this order.

☐ 14. Attach the empty barrel to the minicolumn again.

☐ 15. Add 2 ml of Column Wash Solution (8.3 mM Tris-HCl, pH 7.5, 80 mM potassium acetate, 40 μM EDTA, 55% ethanol) to the syringe barrel. Insert the plunger and gently push the Column Wash Solution through the minicolumn. Discard the solution that flows through the column.

☐ 16. Remove and discard the syringe and plunger. Transfer the minicolumn to a 1.5-ml microcentrifuge tube. Centrifuge at full speed for 2 minutes to dry the resin. Remove the minicolumn from the tube. Discard the tube and liquid.

☐ 17. Transfer the minicolumn to a new, labeled 1.5-ml microcentrifuge tube. Add 50 μl of TE (10 mM Tris-HCl, pH 8, 1 mM EDTA) to the minicolumn. Avoid getting air bubbles between the resin and the TE; place your pipette tip against the inner wall of the minicolumn before slowly expelling the TE. Let sit for 1–2 minutes.

☐ 18. Centrifuge at full speed for 30 seconds to elute the DNA.

☐ 19. Discard the minicolumn. The microcentrifuge tube contains your purified plasmid DNA, which should be stored in the freezer. (Make sure that your tube is labeled.)

DATA ANALYSIS

- Upon completion of Part I of Lab 11C, you will be able to determine whether you successfully isolated plasmid DNA.

LAB 11C

Restriction Cleavage and Mapping of α-Amylase Plasmid DNA

BACKGROUND

Restriction mapping is the process of cleaving DNA with a set of restriction enzymes, both singly and in combination, separating the fragments by gel electrophoresis, and deducing from the fragment sizes the distances between the cleavage sites. A restriction map displays the relative location of the cleavage sites of the set of restriction enzymes. Restriction mapping is essential for many subsequent manipulations of cloned DNA. For example, frequently a smaller fragment of the original cloned DNA needs to be inserted, or subcloned, into another vector. A map of the original cloned DNA is needed to help determine which restriction enzymes will cut out the desired fragment (and these restriction sites must be in the MCR of the new cloning vector).

Recall that restriction enzymes recognize palindromic (inverted repeat) sequences and cleave between specific nucleotides in or near the recognition sequence. Restriction enzymes recognize a sequence of four to eight nucleotides in length. The shorter the recognition sequence, the more likely it is to appear in a DNA molecule. In restriction mapping, it is often desirable to use enzymes that cleave less frequently, thus giving fewer fragments and making the job of deducing the fragment order much easier. Enzymes that recognize a six base-pair sequence are the most commonly used for restriction mapping.

The number of fragments produced after restriction cleavage of plasmid DNA depends upon how many times the circular molecule has been cut. One cut will produce one linear fragment, two cuts will produce two fragments, etc. When two enzymes are used in combination, the total number of fragments should equal the sum of fragments from each enzyme used singly. Constructing a plasmid map is like solving a puzzle. Generally, it is easiest to locate the sites of enzymes that cut the DNA only once. Then the sites for enzymes that cut more frequently can be positioned relative to those sites already mapped.

LABORATORY OVERVIEW

So far, you have identified $amyE^+$ clones by their ability to degrade starch (Lab 10E) and the presence of the 433 bp $amyE$ PCR fragment (Lab 11A). The $amyE^+$ plasmids should contain the 3.3 kb $amyE$ gene fragment, but as discussed in Lab 10C, they could also contain other fragment(s) of *B. licheniformis* DNA. Additional fragments of DNA would make your restriction mapping extremely difficult. You can readily determine whether or not your $amyE^+$ plasmid contains

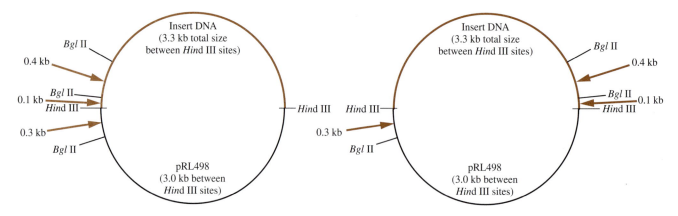

FIGURE 11C.1

Two orientations of 3.3 kb *amyE* gene insert in pRL498. The upper half of the plasmid represents the 3.3 kb *amyE* gene insert, and the lower half of the plasmid represents the 3.0 kb vector, pRL498. Approximate sizes of *Bgl* II/*Hind* III fragments are indicated.

extra inserted DNA by determining the size of the recombinant plasmid. Since you ligated the 3.3 kb *amyE* gene fragment with the 3.0 kb *Hind* III-linearized pRL498, the recombinant plasmid should be 6.3 kb in size.

Since both ends of the 3.3 kb *amyE* gene fragment contained identical *Hind* III sites, it could have been inserted in the 3.0 kb plasmid vector in either of two orientations (Fig. 11C.1). The insert has two cleavage sites for the restriction enzyme *Bgl* II at one end, and the pRL498 plasmid has one *Bgl* II cleavage site. These asymmetric cleavage sites will allow you to determine the orientation of your insert relative to the vector.

In Part I of this lab, you will cleave your *amyE*+ plasmid DNA with *Bgl* II. After gel electrophoresis, you will be able to determine the size of your *amyE*+ plasmid by totaling the size of all the fragments generated by *Bgl* II cleavage. If your *amyE*+ plasmid DNA is of the correct size, then an examination of the sizes of these fragments will reveal the orientation of the 3.3 kb *amyE* gene insert. If the 3.3 kb fragment is oriented as shown on the left of Figure 11C.1, cleavage with *Bgl* II will release a 5.5 kb fragment and two 0.4 kb fragments. If the 3.3 kb fragment is in the opposite orientation, as shown on the right of Figure 11C.1, cleavage with *Bgl* II will release 3.1, 2.8, and 0.4 kb fragments. Lastly, you will estimate the concentration of your plasmid DNA for use in subsequent experiments.

In Part II of this lab, you will cleave your 6.3 kb *amyE*+ plasmid DNA with additional restriction enzymes, singly and in combination, in order to construct a restriction map. The gel containing the fragments of plasmid DNA will also be used for the Southern analysis of Lab 11E.

SAFETY GUIDELINES

Boiling agarose can cause burns. Wear protective gloves when handling hot agarose solutions.

The electric current in a gel electrophoresis device is extremely dangerous. Never remove the lid or touch the buffer once the power is turned on. Make sure that the counter where the gel is being run is dry.

Ethidium bromide is a strong mutagen and a possible carcinogen. Gloves should always be worn when handling gels or buffers containing this chemical.

UV light, used to illuminate the DNA dyed with ethidium bromide, is dangerous. Protect your eyes and face by wearing a UV blocking face shield.

Part I: Determine the Size and Orientation of the Insert

PROCEDURE

CLEAVE amyE⁺ PLASMID DNA INTO FRAGMENTS WITH Bgl II

☐ 1. For each of your samples of *amyE*⁺ plasmid DNA (Lab 11B), label a 1.5-ml tube as "*Bgl* II-digest" (include the clone name, the date, and your group number).

☐ 2. If your *amyE*⁺ plasmid DNA was frozen, thaw it completely, vortex to mix, and centrifuge briefly. Completely thaw the tube of 10× buffer (provided with the enzyme), vortex to mix, centrifuge briefly, and store on ice. Briefly centrifuge the tube of *Bgl* II enzyme, and store on ice. (Review the section on Proper Enzyme Usage in Appendix II.)

☐ 3. Set up each of your *Bgl* II reactions as follows. Add the reagents in the order listed, using a new tip for each reagent. The enzyme is provided in a glycerol storage solution; pipette it slowly and carefully.

dH₂O	15.5 μl
Plasmid DNA	2.0 μl
10× *Bgl* II buffer	2.0 μl
Bgl II	0.5 μl
Total	20.0 μl

☐ 4. For each of your *amyE*⁺ clones, prepare a tube of undigested plasmid DNA that contains the same amount of DNA and buffer, but no *Bgl* II enzyme. Adjust the volume of dH₂O to provide a total volume of 20 μl.

☐ 5. Flick each tube to mix, and spin briefly. Incubate all the tubes at 37°C for at least 4 hours. If the agarose gel will not be run in the same lab session, store the tubes in the freezer until the next lab period.

CHECK CLEAVAGE BY GEL ELECTROPHORESIS

☐ 1. Cast a 0.7% agarose gel as described in Appendix II.
☐ 2. If your samples of *Bgl* II-digested and undigested plasmid DNA were frozen, place the tubes at room temperature and allow them to thaw completely. Mix, and centrifuge briefly.
☐ 3. Add 3 μl of 6× DNA Gel Loading Buffer to each sample. Mix, and centrifuge briefly.
☐ 4. Load the entire volume of each DNA sample in adjacent lanes of the gel. Load DNA molecular size markers on the gel. Keep a record of gel loading.
☐ 5. Run the gel at 80–90 volts until the blue dye has migrated more than halfway down the gel. **WEAR GLOVES WHEN HANDLING ETHIDIUM BROMIDE, AND WEAR EYE PROTECTION WHEN USING UV LIGHT.** Stain the gel with ethidium bromide as described in Appendix II. Photograph the gel; label and place the photo in your notebook.

DATA ANALYSIS

- Determine whether you isolated plasmid DNA in Lab 11B.
- Determine whether *Bgl* II cleaved your plasmid DNA into fragments by comparing the lanes with the digested and undigested DNA. If the plasmid DNA was completely digested, these two lanes should not have bands that migrated the same distance (unless a fragment of linear DNA happens to have the same mobility as the uncut, circular plasmid DNA).
- For each *amyE*⁺ plasmid, estimate the sizes of the fragments produced by *Bgl* II cleavage. The small 0.4 kb fragments may be very difficult to see. Total the sizes of the fragments in each lane of cleaved DNA. Do they add up to the expected size of 6.3 kb? If they add up to more than 6.3 kb, what can you conclude about that *amyE*⁺ plasmid?
- For each *amyE*⁺ plasmid of the correct size, determine the orientation of the 3.3 kb insert.
- For each *amyE*⁺ plasmid of the correct size, estimate the concentration of DNA in the tube of miniprepped plasmid DNA. First, estimate the mass of DNA on the gel by comparing the intensity of ethidium bromide staining of each *Bgl* II fragment relative to that of a DNA size marker band. Second, divide the total mass of DNA by the volume of the plasmid DNA you ran on the gel (you added 2 μl of *amyE*⁺ plasmid DNA to your cleavage reaction, and you loaded the entire volume of the reaction on the gel). Record the concentration of your plasmid DNA in your notebook, and write it on your tube of *amyE*⁺ plasmid DNA.

Part II: Restriction Mapping of α-Amylase Plasmid DNA

To map the 6.3 kb *amyE*⁺ plasmid DNA, it will be cleaved with the following sets of restriction enzymes or enzyme combinations. Each group will choose one of their *amyE*⁺ plasmids (that was shown to be of the correct size in Part I) and then cleave it into fragments with one or more of the following enzymes sets (1–7). If plasmids with both insert orientations are isolated, groups with the same orientation should work together to cleave their DNA with all of the enzyme sets.

Set 1
*Hin*d III alone
*Eco*R I alone
*Hin*d III and *Eco*R I together

Set 2
Pst I alone
*Eco*R I alone
Pst I and *Eco*R I together

Set 3
*Hin*d III alone
Pst I alone
*Hin*d III and *Pst* I together

Set 4
Cla I alone
*Eco*R I alone
Cla I and *Eco*R I together

Set 5
Cla I alone
*Hin*d III alone
Cla I and *Hin*d III together

Set 6
Cla I alone
Pst I alone
Cla I and *Pst* I together

Set 7
Bgl II alone
*Eco*R I alone
Bgl II and *Eco*R I together

Each restriction enzyme is supplied with a stock buffer solution of the optimal pH, ionic strength, and salt composition. This is the buffer to use for single enzyme reactions. Different restriction enzymes require very different reaction conditions, however. Thus, when DNA is to be cleaved with two enzymes simultaneously, you have to determine which buffer will allow each enzyme to function as best as possible. Consult tables in the back of catalogs (e.g., Promega, Inc., New England BioLabs, Inc., etc.) to determine which stock buffer to use for each double enzyme reaction.

PROCEDURE

ASSEMBLE CLEAVAGE REACTIONS

☐ 1. First determine the amount of each reagent to add to your reaction tubes. The volumes of buffer and enzyme to be used are listed in the following chart. Calculate the volume of your 6.3 kb *amyE*$^+$ plasmid DNA that contains about 150 ng of DNA. You determined its concentration of DNA in Part I of this lab. (Refer to the sample calculations on p. 101, if necessary.) Calculate the volume of dH$_2$O needed to bring the total volume of each reaction to 20 μl. Fill in the blanks that follow and record this in your notebook.

	One enzyme	Two enzymes
dH$_2$O	___ μl	___ μl
amyE$^+$ plasmid (~150 ng)	___ μl	___ μl
10× buffer [1× final]	2.0 μl	2.0 μl
Restriction enzyme 1	0.5 μl	0.5 μl
Restriction enzyme 2	0	0.5 μl
Total volume	20.0 μl	20.0 μl

☐ 2. If your tube of *amyE*$^+$ plasmid DNA was frozen, thaw it completely, and centrifuge briefly. Completely thaw the appropriate tubes of 10× buffers. Vortex to mix, centrifuge briefly, and store on ice. Briefly centrifuge the tubes of restriction enzymes, and store on ice. (Review the section on Proper Enzyme Usage in Appendix II.)

☐ 3. Label a 1.5-ml microcentrifuge tube for each of your digestions, including the name of the restriction enzyme(s) to be used, the date, and your group number. Based on your calculations in step 1, add the reagents to each tube in the order listed. Use a new tip for each reagent. The enzyme is provided in a glycerol storage solution; pipette it slowly and carefully.

☐ 4. Prepare a tube of uncleaved control DNA that contains about the same amount of $amyE^+$ plasmid DNA and buffer, but no enzyme. Adjust the volume of dH$_2$O to provide a total volume of 20 μl.

☐ 5. Flick each tube to mix, and centrifuge briefly. Incubate the tubes at 37°C for at least 2 hours. If the agarose gel will not be run in the same lab session, store the tubes in the freezer until the next lab period.

ELECTROPHORESE THE DNA

☐ 1. Cast a 0.7% agarose gel as described in Appendix II.

☐ 2. If your samples were frozen, place the tubes at room temperature and allow them to thaw completely. Mix, and centrifuge briefly.

☐ 3. Add 3 μl of 6× DNA Gel Loading Buffer to each sample, vortex to mix, and centrifuge briefly.

☐ 4. Load the entire volume of each sample in adjacent lanes of the gel. Record in your notebook the order in which you load the samples. It is a good idea to load the samples in alphabetical order so that you can readily recall the order of loading.

☐ 5. Load biotinylated λ DNA/*Hin*d III Markers (for the Southern hybridization of Lab 11E); these markers should be heated to 65°C for 5 minutes before loading on the gel. You can also load unlabeled DNA molecular size markers (e.g., 1 kb DNA ladder).

☐ 6. To save time, you may add ethidium bromide directly to the running buffer (see the section on gel staining in Appendix II). Run the gel at 80–90 volts until the blue dye is about three-fourths of the way down the gel.

☐ 7. ***WEAR GLOVES WHEN HANDLING ETHIDIUM BROMIDE SOLUTIONS, AND WEAR EYE PROTECTION WHEN USING UV LIGHT.*** Carefully remove the gel from the chamber. Cut off the upper right corner of the gel above the wells so that you will know when the gel is right side up. Take a photograph of the gel. Label the photo and place it in your notebook. ***DO NOT DISCARD THE GEL,*** but proceed immediately to Lab 11E.

DATA ANALYSIS

INTERPRET CLEAVAGE RESULTS

• When sufficient plasmid DNA is completely cut into a few fragments of different sizes, interpretation of restriction cleavages is straightforward. Unfortunately, that is rarely the case. Often plasmids are only partially cleaved, and sometimes two or more fragments migrate the same distance, appearing to be a single band. The following guidelines will help you to recognize these problems.

1. Compare the lanes with cut and uncut plasmid DNA. If there is a band in a lane of cleaved DNA that appears virtually identical to the uncut

DNA, you may have had partial cleavage, or one of your fragments may have the same mobility as the uncut plasmid.

2. Add up the sizes of all the fragments. If the total size is more than the expected size of the plasmid (6.3 kb), you probably have partial cleavage products. If the size is much less than you expect, some of the bands may be double bands.

3. Compare the stoichiometry of the bands by their intensity of staining. The staining should be proportional to the size of the band; i.e., a 4-kb band should be twice as bright as a 2-kb band. If some bands seem too bright for their size, they may be double bands. Faint bands are often products of partial digestion. (*Note:* If the gel is overloaded with DNA, all bands may be at saturation and thus very bright; therefore, you cannot judge stoichiometry.)

- Estimate the size of each fragment produced by each of your enzymes or combination of enzymes. For this, you should generate a standard curve that plots the molecular size of the DNA size markers versus the distance each migrated (refer to p. 114). Some of the lanes will have very small fragments that could be smaller than your DNA size markers. Since you will not be able to use your standard curve to estimate the size of these, you will have to estimate size based upon the distance each migrated on the gel. Record the sizes of all the restriction fragments in Table 11C.1. The table should be drawn on the board so that each group can share its results with all members of the class. (You need to know the size of all the restriction fragments to be able to construct a complete map.) Your instructor will help you resolve possible differences in fragment sizes reported by different groups.

CONSTRUCT A RESTRICTION MAP OF α-AMYLASE PLASMID DNA

- Using each set of data from Table 11C.1, draw a map that shows the location of the cleavage site(s) for each restriction enzyme. Begin constructing your restriction map by drawing a simple plasmid map like those shown in Figure 11C.1, in which the restriction sites of the 3.0 kb vector are shown and the 3.3 kb *B. licheniformis* fragment is inserted between the *Hin*d III sites.
- Next map the sites whose location you already know. For example, you know that your *amyE+* plasmid contains two *Hin*d III cleavage sites because cleavage gave two bands. (Also, recall that in Lab 10C, you ligated the 3.3

TABLE 11C.1 Size (kb) of Plasmid Restriction Fragments											
Bgl II	*Cla* I	*Eco*R I	*Hin*d III	*Pst* I	*Bgl* II *Eco*R I	*Cla* I *Eco*R I	*Cla* I *Hin*d III	*Cla* I *Pst* I	*Eco*R I *Hin*d III	*Eco*R I *Pst* I	*Hin*d III *Pst* I
Total:	Total:	Total:	Total:	Total:	Total:	Total:	Total:	Total:	Total:	Total:	Total:

kb *amyE* gene fragment with the 3.0 kb pRL498 vector.) You can also read-ily determine that your *amyE*$^+$ plasmid contains three *Pst* I cleavage sites (because cleavage produced three bands). Examination of the map of plasmid pRL498 in Figure 10B.1 shows that the vector has two *Pst* I sites, which are 2.0 kb apart. This means that the other *Pst* I site lies within the insert, halfway between the two sites in the vector (and cleavage yields two 2.15 kb bands).

- In Part I of this lab, you determined the orientation of your insert on the basis of *Bgl* II cleavage. Now you can use the results of the *Bgl* II/*Eco*R I double cleavage to map the location of the *Eco*R I site relative to the *Bgl* II sites in the insert. Determine the location of the other cleavage sites by examining the sizes of fragments produced by cleavage with two enzymes. The restriction map of a region of *B. licheniformis* DNA that contains the *amyE* gene presented in Figure 9A.1 may be helpful in locating some of the other restriction sites. Remember, however, that your 3.3 kb insert includes about 1.4 kb of additional DNA.

- Combine all of the sites you have mapped into a final map of the *amyE*$^+$ plasmid, noting the distance between each cleavage site.

LAB 11D

Preservation of Recombinant Strains

BACKGROUND

It is important to preserve new recombinant strains for future use. One method for long-term preservation of strains of *E. coli* is to freeze them at a very low temperature ($-80°C$) in the presence of dimethyl sulfoxide (DMSO). DMSO is added to freshly grown cells just prior to freezing to prevent formation of ice crystals, which can cause cell lysis upon thawing. Strains stored in this way can be revived after many years and thus provide a permanent stock of the strain for future applications. A supply of plasmid DNA should also be kept at $-20°C$ in the event that permanent stocks are lost because of freezer malfunction. Plasmid DNA can then be transformed into *E. coli* cells to replace the lost strain.

SAFETY GUIDELINES

Use sterile technique when handling culture media and *E. coli* cells. Discard all contaminated tubes and tips into waste bags, which will be autoclaved prior to disposal.

PROCEDURE

You will prepare a permanent stock of each strain of *E. coli* shown to contain the 6.3 kb *amyE*$^+$ plasmid (Part I of Lab 11C).

INOCULATE CULTURES

1. ☐ ***USE GOOD STERILE TECHNIQUE.*** Label a sterile culture tube for each strain of *E. coli* carrying the 6.3 kb *amyE*$^+$ plasmid. Use a sterile serological pipette to add 2 ml of LB broth + Km to each tube.

2. ☐ Recover the stock plates on which you streaked your putative *amyE*$^+$ colonies in Lab 10E. Use a sterile inoculating loop to transfer one colony from the appropriate stock plate to the corresponding tube.

   Note: If not done already, restreak some of these cells onto a fresh LB agar + Km plate to maintain a constant supply of healthy cells, which are needed for Lab 12. Use a sterile inoculating loop to pick some cells from your original stock plate, and quadrant streak (see Lab 1) over the new LB agar + Km plate. Also, make a new stock plate of the negative cells you picked in Lab 10C, which are also needed for Lab 12.

☐   3.   Incubate the tubes at 37°C overnight while shaking (~225 rpm).
        Note: If the culture will not be used the next day for preparation
        of the permanent stock, store the inoculated broth at 4°C until the
        day before the next lab period, at which time it should be placed in
        the 37°C shaker.

PREPARE PERMANENT STOCKS

☐   1.   For each of your *amyE*⁺ clones, label a sterile 1.5-ml microcen-
        trifuge tube (or a small cryogenic vial).

☐   2.   Aseptically transfer 800 μl of the appropriate *E. coli* culture mixture
        to each tube. Add 200 μl of sterile DMSO to each tube. Vortex to
        mix, and immediately place the tube at −80°C for long-term storage.

LAB 11E

Southern Analysis of α-Amylase Plasmid DNA

LABORATORY OVERVIEW

Restriction analysis yields a map of the recombinant plasmid but does not identify which restriction fragments contain the *amyE* gene. Southern analysis using the BIO-labeled probe from Lab 8 will identify which restriction fragments contain the region of the *amyE* gene that was amplified by PCR.

SAFETY GUIDELINES

Ethidium bromide is a strong mutagen and a possible carcinogen. Gloves should always be worn when handling gels or buffers containing this chemical.

UV light, used to illuminate the DNA stained with ethidium bromide, is dangerous. Protect your eyes and face by wearing a UV-blocking face shield.

The Southern Denaturation Solution contains 0.4 M sodium hydroxide, which is mildly corrosive at this concentration. Wear gloves when handling this solution.

Wear gloves when handling nylon membranes to prevent the transfer of oils, proteins, and other contaminants from your hands to the membrane. Handle the membrane by the edges only and keep it as flat as possible during all the steps because creases, excess pressure, and points of contact will develop increased background staining.

PROCEDURE

DENATURE THE DNA

☐ 1. ***WEAR GLOVES WHEN HANDLING ETHIDIUM BROMIDE SOLUTIONS, AND WEAR EYE PROTECTION WHEN USING UV LIGHT.*** Place your ethidium bromide-stained gel (from Part II of Lab 11C) on the transilluminator and trim away any unused lanes. Cut off the upper right corner of the gel, if necessary.

☐ 2. Make an overlay of the gel that you will use later to help you determine the sizes of the hybridized fragments. Place a piece of clear plastic wrap over the stained gel, smoothing out any wrinkles. Mark the wells and the four edges of the gel on the plastic wrap using a fine-tipped permanent-marking pen. Turn on the UV transilluminator and mark the location of the DNA molecular size markers and the bands of plasmid DNA in the other lanes. Make sure that you mark all of the bands in each lane since some lanes will contain small, faint bands. The sum of the sizes of all the bands in each lane should equal 6.3 kb.

☐ 3. Remove the piece of plastic wrap, rinse the side that touched the gel with water, and blot the wrap dry with a paper towel. This overlay can be traced to make an overlay for the other member(s) of the group. Put this overlay in your notebook.

☐ 4. Place the gel into a shallow container. Add ~100 ml of Southern Denaturation Solution to denature the DNA. Place the container on a rocker or shaker set at low speed and shake gently for 30 minutes.

☐ 5. Remove the container from the rocker and carefully pour off and discard the solution while holding the gel in the container with a gloved hand.

> ⚛ **CAUTION: The denaturation solution makes the gel extremely slippery and fragile!**

☐ 6. Add ~100 ml of Southern Transfer Solution and gently shake or rock the gel for 15 minutes. While waiting, cut your transfer membrane and absorbent paper (see steps 1 and 4 that follow).

TRANSFER THE DNA TO A MEMBRANE

☐ 1. *WEAR GLOVES WHEN HANDLING NYLON MEMBRANES.* Cut a piece of nylon membrane several millimeters larger than your gel. Write your group number or initials near one end of the membrane with a soft lead pencil (do not use ink because it will run during the hybridization procedure). Wet the membrane by placing it in a shallow dish of dH_2O. When the entire membrane is wet, replace the water with ~20 ml of Southern Transfer Solution, submerge the membrane, and let it soak for several minutes.

☐ 2. Set up the Southern transfer, as diagrammed on p. 108. Place a piece of plastic wrap on the counter. Carefully place the gel, right side up, on the plastic wrap. Remember, the upper right-hand corner above the wells was cut off.

☐ 3. Place the wet transfer membrane directly on top of the gel, with your pencil marks facing the gel. Use a glass pipette in a rolling-pin fashion to remove any air bubbles trapped between the gel and the membrane.

☐ 4. Place a piece of thick absorbent paper (e.g., Whatman 3 MM paper), cut slightly larger than the membrane, on top of the nylon membrane.

☐ 5. Place a stack of paper towels, ~1 inch thick, on top of the absorbent paper. Place a glass or plastic plate on top of the paper towels. Place a brick, a book, or other weight on top of the plate. Allow the transfer to proceed for 12–24 hours.

☐ 6. To disassemble the blot, discard the paper towels and absorbent paper. Turn over the gel and transfer membrane so that the gel faces up. Use a soft lead pencil to mark the position of each well through the gel onto the membrane. (You can also mark the edges of the gel on the membrane.) Discard the gel. The DNA is on the side of the membrane that faced the gel and has the pencil marks. (Your blot will be a mirror image of your gel.)

☐ 7. Neutralize the membrane by soaking it in ~20–25 ml of Membrane Neutralization Solution for 3 minutes. Air-dry the membrane and wrap it in foil. (The blue dye on the membrane will disappear during the subsequent hybridization steps.)

AFFIX THE DNA TO THE MEMBRANE

If the DNA will be linked to the membrane with UV light, choose the setting suggested by the manufacturer of the UV cross-linker. Make sure that the DNA side of the blot faces up so that it is fully exposed to the UV light.

If the DNA will be bound to the membrane by baking, place the membrane between two pieces of filter paper and bake it at 80°C for 30 minutes to 2 hours. A vacuum oven is required for baking nitrocellulose membranes but is not required for baking nylon membranes.

HYBRIDIZE THE MEMBRANE WITH BIO-LABELED PROBE DNA

☐   1.   *WEAR GLOVES WHEN HANDLING NYLON MEMBRANES.* Soak the membrane for several minutes in a shallow dish containing ~20 ml of 2× SSC.

☐   2.   Denature the sheared herring sperm DNA by heating at 100°C for 10 minutes. Chill on ice. Add 0.25 ml of the heat-denatured herring sperm DNA to 5 ml of the Southern Prehybridization Solution. Vortex to mix, and keep on ice until ready to use.

☐   3.   Place the wet membrane in a new, quart-sized, zippered freezer bag. Add the Southern Prehybridization Solution (from step 2) to the bag. Remove the trapped air and air bubbles before sealing the bag. If you are using a water bath for hybridization, place the bag with the membrane into a larger freezer bag to prevent water leakage. Remove trapped air from the second bag before sealing.

☐   4.   Incubate at 42°C for at least 1 hour. Gentle agitation is helpful but not essential.

☐   5.   Near the end of the prehybridization period, add the rest (~30 μl) of your BIO-labeled probe from Lab 8 to your (round-bottom, 15-ml tube) tube of recycled Southern Hybridization + BIO-Probe Solution from Lab 9C. Add 100 μl of sheared salmon sperm DNA to the tube of Hybridization + BIO-Probe Solution. Add fresh Southern Hybridization Solution, if necessary, to the tube until the volume reaches 5 ml.

Heat the tube of Southern Hybridization + BIO-Probe Solution at 100°C for 10 minutes to denature the DNA. Immediately place on ice.

☐   6.   Remove your bag containing the membrane from the incubator. Discard the Southern Prehybridization Solution and add the Southern Hybridization Solution + BIO-labeled probe (from step 5). Remove the trapped air and air bubbles before sealing the bag. If you are using a water bath for hybridization, place the hybridization bag into a larger freezer bag to prevent water leakage. Remove trapped air from the larger bag before sealing.

☐   7.   Incubate at 42°C for at least 16 hours. Gentle agitation is helpful but not essential.

Note: After the incubation period, the bag containing the membrane and the hybridization solution can be stored at −20°C for several days.

DETECT THE BIO-LABELED HYBRIDIZED DNA

☐   1.   *WEAR GLOVES WHEN HANDLING NYLON MEMBRANES.* If the membrane and hybridization solution were stored at −20°C, you may have to briefly warm the bag to ~40°C (to bring the salts into solution). Remove the membrane from the bag and place it in a shallow container or a new, quart-sized, zippered bag. Add about 100 ml of Southern Wash Solution 1 to the membrane and rotate gently at room temperature for 3 minutes. If using a zippered bag, remove most of the air bubbles.

2. Drain off the wash solution from the membrane. Add another 100 ml of Southern Wash Solution 1 and rotate at room temperature for 3 minutes.

3. Discard the wash solution and add about 100 ml of Southern Wash Solution 2. Rotate gently at room temperature for 3 minutes. Discard the wash solution. Repeat one time.

4. Discard the wash solution and add about 100 ml of Southern Wash Solution 3 that has been prewarmed to 65°C. Incubate for 10 minutes at 65°C. Repeat one time.

5. Rinse the membrane briefly in about 50 ml of Tris-NaCl. Discard the rinse solution and add about 30 ml of Biotin Blocking Buffer. Incubate at 65°C for 45–60 minutes.

6. Discard the blocking solution and transfer the membrane to a new zippered bag. Add 10 ml of AP-Streptavidin Solution. Incubate at room temperature for 10–15 minutes with gentle agitation.

7. Transfer the membrane to a shallow container and wash in about 200 ml of Tris-NaCl for 5 minutes at room temperature with gentle agitation. Discard the solution. Repeat this wash step three more times.

8. Wash the membrane in about 50 ml of AP Reaction Buffer for 3–5 minutes at room temperature with gentle agitation.

9. Completely drain the reaction buffer and add 20 ml of AP Substrate Solution (BCIP and NBT in AP Reaction Buffer) that was freshly made by your instructor. Allow the reaction to proceed in the dark or in low light. Check the membrane occasionally until blue-purple bands appear. Bands should become visible in 30–60 minutes.

10. Briefly rinse the membrane in about 50 ml of AP Stop Buffer to terminate the color reaction.

11. Photocopy the wet membrane (wrap the membrane in plastic) and place the photocopy in your lab notebook. Wrap the original membrane in aluminum foil to keep light from degrading the color of the bands.

DATA ANALYSIS

- Identify which restriction fragments hybridized with the *amyE* gene PCR probe. To help identify the bands, place the overlay of your gel on your blot. Mark on your overlay those bands that hybridized. Denote (e.g., circle) the hybridizing fragments in Table 11C.1. This table should be drawn on the board again so that all groups can share their data. If some enzymes or enzyme combinations give more than one hybridizing band, what does this indicate to you?

- Map the location of the entire protein-coding region (or ORF) of the *amyE* gene on your plasmid map. It may help to refer to the *B. licheniformis* DNA sequence presented on pp. 83–84 of Lab 7. In Exercise 7.2, you located the start and stop codons of the ORF coding for α-amylase. Also, the *Pst* I cleavage site (CTGCA*G) is located at nucleotide 258 of that DNA sequence.

- Mark on your plasmid map the 433 bp region of the *amyE* gene that hybridized with your probe. (The locations of the PCR primers used to make the BIO-labeled probe were presented in Lab 8.)

QUESTIONS

1. If you were trying to identify a clone that did not have an easily detectable phenotype (as does the *amyE* gene product, which allows the transformed strain of *E. coli* to hydrolyze starch), you would have to use another screening strategy. You would need some way to screen hundreds or thousands of colonies to identify those that contain plasmids with the gene of interest. One technique would be to PCR amplify a portion of the gene directly from the transformed *E. coli* cells (similar to what you did in Lab 11A). Another technique that could be used is called *colony hybridization*. In this technique, transformed *E. coli* cells are grown on (or transferred to) nitrocellulose or nylon membranes, the cells are lysed, the denatured DNA is attached to the membranes, and the membranes are then hybridized with a labeled probe to the gene of interest (similar to the Southern hybridization you did). (For both techniques, you would generate master plates that contained the colonies of cells so that you could retrieve and use them at a later date.)

    a. Which of these two techniques would be *best* if you had to screen a hundred colonies (e.g., using a partial library containing a bacterial gene of interest)?

    b. Which would be *best* if you had to screen a hundred thousand colonies (e.g., using a complete human genomic library)?

    c. What factors (such as cost, reliability, accuracy, etc.) would you consider when deciding which technique would be *best?*

2. Explain how plasmid DNA can be separated from the host cell's chromosomal DNA during plasmid isolation and purification.

3. What additional information does the Southern hybridization provide that the restriction cleavages did not?

4. Suppose that you designed new PCR primers so that you could amplify the entire *amyE* coding region and you used that large BIO-labeled PCR product as a probe with the blot you prepared in this lab. How would your results differ from the results that you obtained using the smaller probe?

5. Identify on Table 11C.1 which additional bands would hybridize with the larger probe (described in question 4) for each enzyme or enzyme combination.

LAB 12

Enzyme Activity of α-Amylase Clones

GOAL

The goal of this lab is to measure α-amylase activity of the *E. coli* strain carrying the cloned *amyE* gene.

OBJECTIVES

After completing Lab 12, you will be able to

1. measure the enzyme activity of a culture of cells that expresses the α-amylase enzyme
2. evaluate the success of the α-amylase cloning project

LABORATORY OVERVIEW

In this lab, you will use the quantitative α-amylase assay (as in Labs 2 and 3) to measure the α-amylase activity of the recombinant strain of *E. coli* carrying the *amyE*⁺ plasmid, negative *E. coli* cells, and *B. licheniformis* cells. You will measure the α-amylase activity inside the bacterial cells and the activity released into the culture medium. To verify that the cloned gene functions like the wild type *B. licheniformis* gene, the enzyme incubation step will be performed at an elevated temperature (100°C).

TIMELINE

This lab requires two lab sessions. Cell cultures are set up during the first session, which takes about 0.5 hour. Enzyme activity is measured during the second session, which takes 1–2 hours.

SAFETY GUIDELINES

Use sterile technique when setting up cultures and handling bacterial cells. Discard all contaminated tubes and tips into waste bags, which will be autoclaved prior to disposal.

Wear gloves when handling the Maltose Color Reagent (1% 3,5-dinitrosalicylic acid, 0.4 M NaOH, 1.06 M sodium potassium tartrate) as this concentration of NaOH is slightly corrosive. Flush your skin or eyes with water if there is contact.

PROCEDURE

INOCULATE CULTURES

1. ***USE GOOD STERILE TECHNIQUE.*** Label three 15-ml sterile culture tubes according to the chart that follows (include your group number or initials). Use a sterile serological pipette to add 3 ml of the appropriate culture medium to each tube.

Tube #	Strain	Medium
1	amyE⁺ E. coli	LB broth + Km
2	Negative E. coli	LB broth + Km
3	B. licheniformis	LB broth

2. Recover the stock plates of your recombinant strain containing the 6.3 kb $amyE^+$ plasmid and the negative strain you picked in Lab 10E (which contains a recombinant plasmid with another chromosomal insert). Use a sterile inoculating loop to transfer one colony from each stock plate to the appropriate tube. Your instructor will provide plates of *B. licheniformis* cells from which you can transfer one colony to the appropriate tube.

3. Incubate the cultures at 37°C overnight while shaking (~225 rpm).
   Note: The inoculated cultures should be grown overnight the day prior to the α-amylase assay. You can store the inoculated broth at 4°C and then place them in the 37°C shaker the day before the assay.

PREPARE SAMPLES FOR ASSAY

1. Vortex the tubes of overnight culture to resuspend the cells. Transfer 1.4 ml of each culture to a new, labeled 1.5-ml microcentrifuge tube.

2. Centrifuge the 1.5-ml tubes for 1 minute to pellet the cells. Transfer all of the culture medium above each pellet to a new, labeled, 1.5-ml microcentrifuge tube. Store the tubes of culture medium on ice while you prepare the cell lysates.

3. Add 0.1 ml of 1× TBS (25 mM Tris-HCl, pH 7.5, 137 mM NaCl, 2.7 mM KCl) to each tube of cells. Vortex vigorously to resuspend the pellet of cells. Check to see that you have homogeneous cell suspensions. Turn each tube upside down and flick it with your finger to disperse the mixture along the walls of the tube. Look to see if there are any clumps of cells. If there are, continue to vortex until the cell mixture is completely homogeneous. Store these tubes on ice.

4. The cells will be pulverized with glass beads to break them open. Add 0.1 ml of glass beads to each of three new, labeled 1.5-ml microcentrifuge tubes, as described on p. 50 of Lab 5A.

5. Use a micropipette to transfer the resuspended cell pellets to the corresponding tube with the beads.

6. Close the lids securely and vortex the tubes of cells and beads for 1 minute. Place the tubes on ice for 1 minute. Repeat this sequence four more times to give a total vortexing time of 5 minutes.

7. Add 1.3 ml of 1× TBS to each tube and vortex to mix. (Now the volume of each lysate equals the volume of starting cells.) Spin the

	TABLE 12.1 Assay Components				
Sample	Condition	LB Broth	Culture Medium[1]	1× TBS	Lysate[2]
1	Culture medium blank	1.0 ml	0		
2	*amyE*$^+$ *E. coli* medium	0.9 ml	0.1 ml		
3	Negative *E. coli* medium	0	1.0 ml		
4	*B. licheniformis* medium	0	1.0 ml		
5	Cell lysate blank			1.0 ml	0
6	*amyE*$^+$ *E. coli* lysate			0.9 ml	0.1 ml
7	Negative *E. coli* lysate			0	1.0 ml
8	*B. licheniformis* lysate				1.0 ml

[1]From step 2.
[2]From step 7. Remove the lysate from the top of the tube without disturbing the pellet at the bottom.

       tubes of cells at maximum speed for 2 minutes. Keep these tubes on ice while you set up the assay tubes.

8. Label eight round-bottom, 15-ml assay tubes, according to Table 12.1.
9. Add the appropriate volume of each reagent, listed in Table 12.1, to each assay tube.

MEASURE α-AMYLASE ACTIVITY

1. Add 1 ml of 1% starch solution, pH 7, to each of the assay tubes, including the blank tubes. Cap, vortex, and place in a heating block set at 100°C (or a boiling water bath) for exactly 12 minutes. Start timing when starch is added to the first tube. During this enzyme incubation period, the heat-stable *B. licheniformis* α-amylase breaks down starch into maltose and other products.

2. In the same order that the starch solution was added, add 1 ml of the Maltose Color Reagent to each tube. ***WEAR GLOVES WHEN HANDLING THE MALTOSE COLOR REAGENT.*** Cap, vortex, and place in a heating block set at 100°C (or a boiling water bath) for exactly 15 minutes. Start timing when the Maltose Color Reagent is added to the first tube. During this step, the Maltose Color Reagent reacts with the maltose that was produced during the enzyme incubation step.

3. Place tubes on ice until cooled to room temperature. Add 9 ml of dH$_2$O to each tube. Tightly cap the tubes and invert several times to mix.

4. Use the appropriate blank to zero the spectrophotometer at 540 nm and measure the A_{540nm} of the samples. Check to make sure each reading is within the range of the maltose standard curve from Lab 2. If it is too high, dilute that sample (and the blank) and measure them again. Record the absorption readings in Table 12.2.

DATA ANALYSIS

- Use the maltose standard curve you constructed in Lab 2 to determine the amount of maltose present in each sample tested. Record these in Table 12.2.
- Calculate the amount of maltose that would have been produced by 1 ml of each test solution, and record these values in Table 12.2. *Note:* The negative *E. coli* and the *B. licheniformis* assay tubes contained 1 ml of test solution, whereas the *amyE*$^+$ *E. coli* assay tubes contained only 0.1 ml of test solution. Place this table in your notebook.

	$amyE^+$ E. coli Medium	Negative E. coli Medium	B. licheniformis Medium	$amyE^+$ E. coli Lysate	Negative E. coli Lysate	B. licheniformis Lysate
TABLE 12.2 — α-Amylase Activity of Bacterial Cell Cultures						
A_{540nm}						
mg of maltose produced during assay						
Calculated mg of maltose produced by 1 ml of test solution						

- Add together the activity in the culture medium and the lysate for each strain to determine the total α-amylase activity in each. Which strain has the highest α-amylase activity?

QUESTIONS

1. How does the amount of α-amylase activity compare between the recombinant strain of *E. coli* with the *amyE* gene and *B. licheniformis*, the source of the gene? Does the location of the enzyme (i.e., released into the cell medium versus present in the cells) differ between the two strains? Are both enzymes heat stable?
2. Explain why the *E. coli* cells carrying the 6.3 kb *amyE+* plasmid are able to produce more α-amylase enzyme than *B. licheniformis*.

EXERCISES

EXERCISE 12.1
Evaluate Your Hypothesis

Review your hypothesis from the beginning of the project. Did your hypothesis prove to be scientifically correct? Explain how your results support your conclusion.

EXERCISE 12.2
Course Wrap-up

Suppose you wanted to sell your recombinant strain of *E. coli* that expresses the *amyE* gene to a company that purifies α-amylase from *B. licheniformis* for the conversion of cornstarch to corn syrup. Write a one-page scientific proposal to them explaining why this clone would increase their profits. Explain the advantages and possible disadvantages of using your *E. coli* clone versus *B. licheniformis*. Some things to consider are the amount of α-amylase produced per volume of cells, the ease of purifying the α-amylase enzyme from either strain, etc. (You can assume that the cost to grow *E. coli* and *Bacillus* cells would be about the same.) If you do not believe that this clone would increase their profits, explain why this project may have been an economic failure.

Appendix I

Additional Information and Exercises

BIOINFORMATICS: AN INTRODUCTION

Bioinformatics is a relatively new field that integrates biology with information science. As the number and size of databases of biochemical and genetic information have increased over the last few years, it has become important to organize the information and to develop software to understand the significance of the information. This requires the use of computer systems and the development of an interface that makes the data available and usable to scientists all over the world.

Bioinformatics encompasses many aspects of biology and chemistry, from small chemical molecules to large genomes such as the human genome that contains 3 billion chemical bases. The sequencing of the genomes from many bacteria, plants, and animals has produced vast amounts of information that is useless until it is organized and analyzed. The algorithms that have been developed for analysis of complex biological information form the basis for bioinformatic studies. While it is beyond the scope of this introduction to bioinformatics to explain many aspects of bioinformatics in detail, it is possible to give brief descriptions of some of the important techniques and to give links to websites that provide more information. The discussion here will be restricted to DNA and proteins; however, you should understand that other molecules can also be analyzed.

DNA Sequences

The field of bioinformatics has developed as the techniques for DNA sequencing have evolved. When DNA sequencing techniques were first developed, only a few hundred nucleotides could be determined in a day. With newer sequencing methods and automation, millions of nucleotides of preliminary sequence can be obtained in about the same amount of time. While production of the sequence data is now quite rapid, the problem is ensuring the accuracy of the data and developing methods for rapid analysis of such vast amounts of information. Computers have become the storehouses for all the sequence information. International databases such as GenBank and EMBL store the information and allow the sequences in the databases to be searched, analyzed, and even downloaded to individual computers.

Open Reading Frames

The DNA sequence is a code that must be translated both in the cell and by the computer to provide information. Just as the cell uses molecular machinery to translate some DNA into protein, computer programs can translate DNA sequence into amino acids. The problem is that simple translation programs cannot know which reading frame encodes an active protein or whether a particular region of sequence actually encodes a protein. Using certain rules, such as the translated sequence must begin with a methionine (typically the first amino acid in a protein), must be at least 100 amino acids long, and must end with one of the three termination signals, computer programs can identify possible protein coding regions, called *open reading frames* (ORF). However, the program cannot determine whether an ORF is actually used by a cell to encode a protein, and it will miss those genes that do not conform to the rules. Nevertheless, the ability to easily identify all ORFs in a complex genome is an important first step to understanding function.

Identification of Proteins

Each ORF can be further analyzed by comparing the predicted amino acid sequence to other known proteins in other organisms. Because proteins with similar function often have conserved amino acids sequences, it is often possible to identify the probable function of an ORF if it has a high degree of similarity to another protein whose function has been determined. As more members of a family of conserved proteins are identified, it becomes easier to identify new members by their similarity to all the other members of the family. This can be done using programs for sequence alignment that allow many members of a family of related proteins to be compared directly with each other. Once a good alignment of several proteins has been obtained, other programs can determine which proteins are more related to each other and which are less closely related. From this information, still other programs produce phylogenetic trees that show the relationships of all the proteins to each other.

Other DNA Sequence Information

Not all DNA encodes proteins. Some sequences are not used at all, other regions are important for regulation, and still other regions encode ribosomal RNA and transfer RNA molecules. Because of their similarity across all organisms, RNA molecules are readily identified from the DNA sequence. Regulatory regions of DNA are much more difficult to identify because they are often not well conserved among organisms. The lack of any reasonable ORF or evidence of other function is often the criterion that identifies noncoding regions. The function of the vast amounts of noncoding DNA in higher organisms remains to be determined.

Proteomics

The progress in determining genome sequences has been paralleled by similar progress in understanding the structure and function of proteins. Proteins can be genetically engineered to be very highly expressed in bacterial cells, allowing scientists to purify amounts sufficient for detailed analysis of structure. The three-dimensional structure of thousands of proteins has been determined by X-ray crystallography, and that information is stored in the Protein Data Bank, where it is freely available to the public. In addition, programs such as RasMol render 3-D images of the proteins that are easily seen and manipulated on any computer. More complex programs can analyze the structure, folding pattern, and chemical interactions to allow scientists to begin to understand how the protein functions.

Certain amino acid sequences that are highly conserved in all proteins that share a particular function allow identification of a particular motif. For example, many DNA binding proteins contain an amino acid sequence that folds into a structure called a *helix-turn-helix*. This means that the protein has an α-helix followed by a turn in the strand, followed by another α-helix. When such a motif is identified, this suggests that the protein may bind to DNA and may be involved in gene regulation. Computer programs can analyze proteins to identify many known protein motifs. As the 3-D structure and function of more proteins are determined, more motifs are identified, and these aid in the identification of possible protein function.

Genetics

DNA and protein sequence information provide evidence of function, but genetic studies are often used to determine or to verify function. Mutants can be engineered based on the DNA sequence information. Structural studies of proteins provide information concerning which amino acids are likely to be important for function so that mutants can be made that will affect the protein function. Genetic studies are done in the organisms, bringing together the sequence information and the biology of the organism.

Microarrays

A powerful technique for determining if and when a gene is expressed is Northern analysis. This technique takes advantage of the fact that the messenger RNA that is made when a gene is expressed can be isolated from cells and can then, on a solid surface, form a hybrid with the DNA that encodes the messenger RNA. DNA for a gene of interest will form a hybrid with the messenger RNA for that gene only if that gene of interest is expressed. This technique has been used for many years to determine the conditions under which a gene is expressed. With the sequence of whole genomes now available, it is possible to ask what genes in the entire genome are expressed under a particular set of conditions. Each ORF in the genome is separately amplified and bound as a minute spot to a slide in an organized array of up to 20,000 separate genes. This is called a *microarray*. RNA from cells is labeled with a fluorescent dye and is hybridized to all the DNA sequences on the microarray slide. The RNA that was made from a particular gene matches that gene sequence and will bind to the spot, giving rise to a fluorescent signal at that spot. Since the gene for each spot is known, those genes that fluoresce indicate a gene that is expressed. This technique can be used to identify genes that are expressed only under one particular physiological condition. RNA is extracted from cells under different physiological conditions, and each RNA sample is labeled with a different-colored fluorescent dye. The RNA samples are hybridized to the microarray. A DNA spot may bind RNA from cells grown under one, both, or neither condition. Laser detectors scan the slide and quickly identify by the fluorescent dye which genes are expressed under only one physiological condition. This is a rapid and powerful technique for identifying genes that are important in a particular cell type or a particular growth condition.

More Information

This brief overview of the field of bioinformatics does not begin to describe the many facets of this emerging field. You will find more information by exploring the Web. Some sites that currently provide information and links to many other sites are provided as follows:

<http://bioinformatics.org> (information and software for bioinformatics)

<http://www.bioinformatics.org/SMS> (this site gives examples of the many programs that are used in DNA and protein analysis)

<http://www.nwfsc.noaa.gov/bioinformatics.html> (great source for links to many resources)

<http://www.ncbi.nlm.nih.gov/> (NIH site for DNA sequence files and much other information)

<http://www.rcsb.org/pdb/> (Protein Data Bank)

<http://www.expasy.ch/> (protein analysis)

<http://www.umass.edu/microbio/rasmol/> (protein structure visualization programs)

<http://twod.med.harvard.edu/seqanal/> (sequence comparisons)

<http://www.blocks.fhcrc.org/> (protein sequence alignment)

<http://www.motif.genome.ad.jp/> (for finding DNA or protein motifs)

BASIC BLAST SEARCH

GOAL

The goal of this exercise is to use BLAST (Basic Local Alignment Search Tool) to identify an unknown sequence.

OBJECTIVES

After completion of this exercise, you will be able to

1. access the BLAST server via the NCBI (National Center for Biotechnology Information) at NIH (National Institutes of Health)
2. choose the correct BLAST search tool to compare an unknown sequence to sequences in the databases at NCBI
3. identify the probable sequence from the sequence data provided

BACKGROUND

The National Center for Biotechnology Information (NCBI) in the National Library of Medicine (NLM) at the National Institutes of Health (NIH) maintains many DNA and protein sequence databases. These sequences are submitted by scientists to the various databases, and NCBI makes these data available to the public over the Internet. Each sequence in a database has an accession number, information relevant to the sequence, and the sequence itself.

The computer program that is used to search these databases is called BLAST (Basic Local Alignment Search Tool), which has been developed by Altschul and Schuler (Schuler et al. 1991; Altschul et al. 1994). There are BLAST programs for aligning either DNA sequences or protein sequences. The blastn program searches the databases for DNA sequences similar to the DNA sequence of interest, which is called the *query* sequence. The blastp program searches the databases for amino acid sequences similar to the amino acid query sequence. The blastx program compares a nucleotide query sequence, translated in all six reading frames (three in the forward direction and three in the reverse direction), against a protein sequence database. For a more complete discussion of BLAST programs, see the following website:

<http://www.cshl.org/books/g_a/bk1ch7/seqanl.html#blast>

PROCEDURE

Assume you have sequenced a fragment of DNA and used a computer program to translate that DNA sequence into a sequence of amino acids. These two query sequences are provided as follows and will be used in the following exercises to allow you to determine the probable identity of the gene that you sequenced.

BLAST Search Query DNA Sequence

atgaagttgt ttctgctgct ttcagcccttt gggttctgct gggcccagta tgccccacaa
acccagtctg gacgaacgtc tattgtccat ctgtttgaat ggcgctgggt tgacattgct
cttgaatgtg agcggtattt gggcccaaag ggatttggag gggtacaggt ctcccccccc
aatgaaaaata tagtagtcac taacccttca agaccttggt gggagagata ccaaccagtg
agttacaagt tatgtaccag atcaggaaat gaaaatgaat tcagagacat ggtgactaga
tgtaacaacg ttggcgtccg tatatatgtg gacgctgtca ttaatcatat gtgtggaagt

BLAST Search Query Amino Acid Sequence

mklflllsafgfcwaqyapqtqsgrtsivhlfewrwvdialecerylgpkgfggvqvsppn
enivvtnpsrpwweryqpvsyklctrsgnenefrdmvtrcnnvgvriyvdavinhmcgs

Search with blastn

- Go to <http://www.mhhe.com/thiel> (the website for this manual). Find the section on Bioinformatics and scroll down to the section entitled "BLAST Search Query DNA Sequence." Using the mouse, select only the DNA sequence and copy it to the clipboard of your computer.
- Go to the NCBI website at <http://www.ncbi.nlm.nih.gov/> and select "BLAST" from the menu line at the top of the screen.
- Select "Standard nucleotide-nucleotide BLAST [blastn]" under the "Nucleotide BLAST" option and a new screen will appear.
- Paste your query DNA sequence (from your clipboard) into the big "Search" box. Use the default settings, i.e., you want "nr" in the "Choose database" box, and further down the page, under the "Options for advanced blasting" section, the "low complexity" box should be checked for "Choosing filters." Click on the "BLAST!" button to start the search.
- A new screen will appear that states "Your request has been successfully submitted and put into the Blast Queue." Below the line that reads "The request ID is xxx-xxx" click on the "Format!" button. The "BLAST Search Results" screen will appear next. When the search is completed, the results will be displayed. The search may take a few seconds or a few minutes, depending on the time of day.
- Scroll down to "Distribution of # Blast Hits on the Query Sequence." This shows you a graphical view of all the sequences that are similar to the query sequence. They are color-coded bars, with red indicating sequences that are very similar and black indicating sequences that are less similar. If you place the mouse over a colored bar, the name of the sequence will appear in the box above the colored bars. The name may include information about the source of the DNA (what organism and what tissue) and the identity of the sequence (name of gene). If you click on a colored bar, it will take you to the alignment of that sequence with the query sequence. This alignment is presented further down the page. To get back to the top of the page, click on the "Back" button of your browser or simply scroll back up to the top.
- Scroll down past the box with the colored bars to a list of the "Sequences producing significant alignments." On the left side of the list are links that you can click on that will take you to the sequence file in the database. Both the name of the database (gb = GenBank, emb = European Molecular Biology, etc.) and the accession number of the sequence are presented. In the sequence file, you will find the name of the gene, the source of the DNA, the title of relevant journal articles, the translated protein sequence, the DNA sequence, and much more information. Frequently, the titles of the journal reveal the name of the gene and the organism (and tissue) from which the DNA was derived.
- On the right-hand side of the list, there is a "Score" and an "E value" for each sequence. The score indicates the degree of similarity between that sequence and the query sequence; the higher the score, the better the match. The lower the E value, the higher the probability that the match between the two sequences is not due to random chance. The E value for good matches is usually expressed as a negative exponent (e.g., e^{-134}), which means that there is a very low probability that the sequences match only by random chance. An E value of 0 indicates identity or near identity of the database sequence with the query sequence. Short sequences of identical matches

give higher E values than longer sequences of identical matches because there is a greater probability of shorter sequences matching by chance than of longer sequences matching by chance. *Note:* there may be sequences in this list that are not presented graphically as colored bars, in the top part of the screen.

- Below the list of similar sequences you will find the alignment of each of the database sequences to the query sequence itself. Below the name of each sequence, the similarity scores and E values are presented, as well as the percentage of identity between the two DNA sequences.

1. What is the identity (name of the gene or protein encoded by the gene) of the database sequence that is most similar to the query sequence, and how similar is it to the query sequence? _____
_____

2. What is the source (organism and tissue) of the database sequence that is most similar to the query sequence? _____
_____

3. For the next three most similar sequences, mouse-over or click on the colored bars to identify the gene, the source (organism and tissue) of the sequence, and the % nucleotide identity between that sequence and the query sequence.

	Name of gene	Source	% Identity
Second:	_____	_____	_____
Third:	_____	_____	_____
Fourth:	_____	_____	_____

4. What gene did you sequence? (In other words, what is the identity and source of your query sequence?) _____
_____

Search with blastp

- Go to the website for this manual. Select and copy the "BLAST Search Query Amino Acid Sequence." Repeat the steps of the previous exercise using the amino acid sequence as the query sequence to search the databases for similar amino acid sequences. This time, select "Standard protein-protein BLAST [blastp]" under the "Protein BLAST" option. Paste your amino acid sequence into the "Search" box; keep "nr" as the database but remove the check from the "low complexity" filter box. You do not want to filter out low complexity regions.
- When the results of your protein database search are displayed, the first thing you should notice is that there are many more red bars presented than in your earlier nucleotide database search.

5. Why does an amino acid query sequence result in higher similarity scores than a nucleotide query sequence? (Think about the genetic code.)
_____
_____
_____

6. Identify the four database sequences that are most similar to the query sequence. List the name of the protein, the source (organism and tissue) of the protein, and the % of amino acid identity.

   In the alignment section, there may be several sequences (with their links to the database) listed together because they are all *identical*. These sequences were submitted to the databases by different researchers.

Below the names of the sequences, the similarity scores, E-values, % identities, and % positives (i.e., similar amino acids) are presented.

	Name	Source	% Identity
First:	_____	_____	_____
Second:	_____	_____	_____
Third:	_____	_____	_____
Fourth:	_____	_____	_____

7. What animals, besides the ones listed previously, have this protein and have a similarity score of greater than 230 bits? _____
    _____

Search with blastx

You can repeat the preceding steps using the blastx program to search the databases for amino acid sequences that are similar to all six reading frames of the translated DNA query sequence, i.e., your downloaded DNA sequence. This time, select "Nucleotide query–protein db [blastx]" under the "Translated BLAST Searches" option. Search the "nr" database and remove the check from the "low complexity" filter box. Does this search provide the same results as the previous searches?

SEQUENCE ANALYSIS USING COMPUTER SOFTWARE

GOAL

The goal of this exercise is to use a computer program to analyze the α-amylase gene of *B. licheniformis.*

OBJECTIVES

After completing this exercise, you will be able to

1. obtain a DNA sequence from GenBank using the "Search" function of Entrez at NCBI
2. use computer software to translate the DNA sequence into at least three reading frames
3. identify the open reading frame (ORF) most likely to correspond to the gene
4. use BLAST to verify the correct identification of the ORF
5. use computer software to construct a restriction map of the gene

BACKGROUND

The availability of DNA sequences in public databases and the development of software to analyze those sequences have greatly simplified the process of obtaining useful information from DNA sequences. Imagine how tedious it would be to manually translate each possible three-letter code into the corresponding amino acid using just your eyes, a codon table, and a pencil. Similarly, imagine searching a sequence for every instance of a specific sequence of nucleotides that identifies a restriction site. Fortunately, software quickly does this for you.

For this exercise, you must have DNA/protein analysis software. Many institutions own commercial products such as GCG, DNAStar, DNAsis, Vector NTI Suite, DNA Strider, or other similar programs. If such a program is available, you should carefully read the instructions for that particular program in following the exercises that follow. If your institution does not have such software, two websites provide programs for DNA analysis:

Bionavigator at <http://www.bionavigator.com/>
(Many programs are available for use on this website. You must set up an account. At this time, new users are allowed 100 units of computer time on the site for free.)
Jellyfish at <http://www.biowire.com/>
(Look on the *biowire* home page for information about Jellyfish, which is distributed by *biowire*.)

PROCEDURE

Obtain a DNA Sequence

- Using the Web, access Entrez at NCBI at <http://www.ncbi.nlm.nih.gov/Entrez/>. Click on "Nucleotide" to access the nucleotide database.
- Search the nucleotide database for the *Bacillus licheniformis* α-amylase gene (accession # X03236). Type the accession number in the box, and then press the "Go" button.
- Open the sequence file X03236. (Click on the number to go to the sequence file.)

1. How many nucleotides are present in the sequence with accession number X03236? _____

2. Look at the "Features" section of the file. The region of the sequence that encodes the α-amylase precursor is indicated by "CDS." What are the numbers of the starting and ending nucleotides for the sequence that encodes the α-amylase precursor? _____

   _____

3. Part of the precursor protein is a signal peptide that is cleaved as the protein is secreted from the cell. After the signal peptide is removed, the protein is referred to as the mature protein.

   a. Using the "Features" section of the file, identify the part of the gene that encodes the signal peptide (identified as "sig peptide"). _____

   _____

   b. Using the "Features" section of the file, identify the part of the gene that encodes the mature peptide (identified as "mat peptide"). ____

   _____

- Scroll to the bottom of the file until you see the whole nucleotide sequence.
- Select the sequence and copy it to the clipboard. Do not be concerned about the numbers; they should be filtered out by the software.

Analyze the Sequence

- Open the DNA analysis software program you are using. Open a new DNA file.
- Place the cursor in the area for the sequence and paste the α-amylase gene sequence into the space.
- Save the sequence file with an appropriate name, such as "Blichamy."
- Using the appropriate program of your software package, translate the DNA sequence. You should have the option of translating in at least the three forward reading frames. Some programs allow you to view all three translations on the same screen. Other programs show only one reading frame at a time. Most programs will identify the open reading frames (ORF) in each translated reading frame.
- Note the lengths of the ORFs for each of the three forward reading frames and choose the reading frame that gives the longest ORF, which is likely to be the α-amylase gene.

4. Do all three forward reading frames of the DNA sequence contain ORFs? List all the ORFs longer than 50 amino acids. _____

   _____

5. The longest ORF corresponds to the coding region of the α-amylase gene. How long is this ORF? _____

   _____

Identify the ORF

- Select the amino acids in this long ORF and copy this amino acid sequence to the clipboard.
- Using the instructions in the Basic BLAST Search exercise, do a BLAST search with this sequence using the blastp program. If everything was done correctly, the BLAST search results should indicate that the protein is α-amylase from *B. licheniformis*.

6. Does the BLAST search indicate that this ORF corresponds to the gene that you used for this analysis? _____

   _____

a. Do any other genes also correspond to this sequence? _____

b. How similar are these other genes? _____

_____

Locate and Map Restriction Enzyme Cleavage Sites

- Go back to the DNA sequence for α-amylase in your DNA analysis program. Use the appropriate program of your software package to identify the location of restriction cleavage sites of the following enzymes: *Cla* I, *Hinc* II, *Kpn* I, *Pst* I, and *Sal* I.

7. List the location of these cleavage sites in the α-amylase gene.

_____

_____

8. Using the positions of the cleavage sites for these enzymes, construct a simple restriction map of the α-amylase gene. (*Note:* Most programs will construct the map for you, and you can simply print the map.)

MICROPIPETTING EXERCISE

GOAL

The goal of this exercise is to test the accuracy and reproducibility of your micropipetting skills.

OBJECTIVES

After completion of this exercise, you will be able to

1. choose the appropriate micropipette to measure a particular volume
2. distinguish between the first position and the second position of the plunger of a micropipette
3. accurately measure and dispense small volumes of liquid

BACKGROUND

One important tool in biotechnology is the adjustable-volume micropipette, which accurately measures and dispenses small volumes of liquid. Although micropipettes are precise tools, their accuracy is very dependent on the skill of the user. The accurate use of a micropipette requires instruction and practice.

Adjustable-volume micropipettes are designed to measure ranges of specific volumes. Generally, they are provided in sets: one to measure from 1–20 μl, another to measure 20–200 μl, and another to measure 200–1000 μl. Learn to distinguish the micropipettes by the color of their knobs or their labels so that you always use the appropriate micropipette.

The volume is set with an adjustment knob and is displayed on a digital indicator. Your instructor will explain how to read the indicators of the micropipettes you will be using. Different manufacturers use different conventions to indicate a decimal place (e.g., the color of the digits switches from black to red, or there is a red line between two digits). Furthermore, the location of the decimal place generally differs between small (1–20 μl) pipettes and large (0.2–1.0 ml) pipettes. These pipettes are designed to work within a specific range of volumes. Outside of that range, they are not accurate or reliable. Thus, *NEVER SET THE DIAL ABOVE OR BELOW THE WORKING RANGE OF THE PIPETTE.*

The most important design feature of an automatic micropipette is that the plunger has two positions. Set the dial to a midrange value, and slowly depress the plunger until you feel a slight resistance. This is the first stop that is used to draw up the liquid. Depress the plunger further to the second stop. It takes more force to reach this position, which is used to dispense the liquid.

The following steps guide you through the proper use of a micropipette:

1. Set the desired volume by turning the volume adjustment knob until the correct volume shows on the digital indicator.
2. Attach a pipette tip to the shaft of the micropipette, making sure the tip fits snugly.
3. Depress the plunger to the first stop.
4. Holding the micropipette at a minimal angle ($<20°$ from vertical), immerse the pipette tip into the sample liquid. Hold the bottle or tube at eye level so you can see what you are doing. The pipette tip should be placed just below the surface of the liquid.
5. Slowly release the plunger until it stops. Watch to see that the pipette tip fills with solution; keep the tip immersed for several seconds to insure that the full volume of the sample is drawn into the tip (this is especially

important with viscous solutions). Withdraw the tip from the sample liquid. If you are pipetting from a microcentrifuge tube, drag the tip up the inner wall of the tube to remove liquid clinging to the outside of the tip.

6. Holding the receiving tube at eye level, place the pipette tip against the wall and slowly depress the plunger to the first stop. Then depress the plunger to the second stop, while dragging the tip up the wall of the tube. This will expel any residual liquid from the tip. Always check that your sample was transferred to the side of the receiving tube. This is particularly important when you are working with small volumes.

7. Once the pipette tip nears the top of the tube, slowly release the plunger. Discard the tip.

OVERVIEW

In the following exercise, you will pipette increasing volumes of a concentrated dye (Fast Green at 0.5 μg/μl) into cuvettes, add water, and then measure the absorbance of each sample. If you measured the dye accurately, a graph of the amount of dye versus absorbance should give a straight line. Each sample will be prepared in duplicate so that you can evaluate the reproducibility of your pipetting.

Each student should set up and read the absorbance of duplicate samples and generate a standard curve. Do not work in pairs or groups.

PROCEDURE

1. In column 1 of the chart that follows, the volume of a solution of Fast Green to be added to each cuvette is shown. Calculate the amount of water needed to give a final volume of 1 ml (or 1000 μl) and write that value in column 2. Calculate the mass (μg) of Fast Green in each sample and write that value in column 4. Mass can be calculated by multiplying the concentration by the volume. For example,

$$0.5 \ \mu g/\mu l \times 40 \ \mu l = 20 \ \mu g$$

Volume of Fast Green (0.5 μg/μl)	Volume of Water Needed (ml)	Total Volume	Mass of Fast Green in Cuvette (μg)
0 μl	1.0 ml	1.0 ml	0
2 μl	998 μl	1.0 ml	
5 μl	995 μl	1.0 ml	
10 μl	990 μl	1.0 ml	
20 μl	980 μl	1.0 ml	

2. Obtain eight 1.6-ml plastic cuvettes, two for each of the different amounts of Fast Green (label each duplicate pair "A" and "B"). Using the appropriate-sized micropipette, add the volume of Fast Green shown in column 1 to each of the duplicate cuvettes.

3. Using the appropriate-sized micropipette (different from the one used in step 2), add the volume of water shown in column 2 to each of the duplicate cuvettes.

4. Prepare a blank for the spectrophotometer by adding 1.0 ml of water to another cuvette.

5. Place a piece of Parafilm over each cuvette, and invert three to four times to mix thoroughly.

☐   6.   Using the cuvette with water as the blank to zero the spectrophotometer, read the absorbance at 620 nm of each sample. Record your values in the following chart. In column 1, copy the mass of Fast Green in each sample that you calculated from the preceding chart (column 4).

μg of Fast Green	A_{620} Reading of Sample A	A_{620} Reading of Sample B	Mean Value of Duplicate Samples	Standard Deviation

DATA ANALYSIS

- Plot your data on graph paper, putting the mass (μg) of Fast Green on the x-axis and the mean A_{620} value for the two samples on the y-axis. Connect the data points.
- If your micropipetting technique was good, the graph you produced should be a straight line. The more points that deviate from a line, the more inaccurate your pipetting skills.
- The standard deviation for the two values of the duplicate tubes provides a measure of the reproducibility of your pipetting skills. The smaller the standard deviation, the better your skills.

Appendix II

Frequently Used Procedures

AGAROSE GEL ELECTROPHORESIS

Mini horizontal agarose gels (e.g., 8 × 10 cm) work well for all the procedures presented in this manual. The number of lanes needed per lab activity varies, but the wells need to hold from 10–25 μl. Generally, each group will pour and run its own gel.

Casting Gels

1. To make a 0.7% agarose gel, weigh out 0.35 g of electrophoresis grade agarose and place it in a 125-ml glass Erlenmeyer flask. (To make a 1.0% gel, weigh out 0.5 g of agarose.) Add 50 ml 1× Tris Acetate EDTA (TAE) Buffer (40 mM Tris-acetate, 1 mM EDTA). Do not swirl the flask as it is extremely difficult to melt agarose that is stuck to the glass above the buffer. Mark the level of the liquid on the outside of the flask with a marking pen. Loosely plug the opening of the flask with Kimwipes to prevent excessive evaporation.

2. Heat the mixture in a microwave oven (or a boiling water bath or a hot plate) until the agarose is completely melted. Wearing heat-protective gloves, carefully swirl the flask to make sure that there are no undissolved grains of agarose floating in the solution. Check the volume of the solution after heating, and add dH$_2$O until the level of liquid reaches your marking on the flask, if necessary. Allow the agarose solution to cool to about 60°C (it is cool enough when you can handle the flask comfortably without protective gloves).

3. Follow the directions provided with your electrophoresis system for assembling your gel-casting unit. If the depth of the comb can be adjusted, position it 1–2 mm from the bottom of the gel tray/chamber. To do this, place two standard glass microscope slides (stacked flat) at one end of the casting tray and adjust the teeth of the comb so they sit on the slides. (This step also ensures that the depth of the comb is uniform across the gel.) Remove the slides and comb before pouring the gel.

4. Pour the slightly cooled melted agarose into the casting tray (do not pour very hot agarose into the casting tray as it will damage the plastic or plexiglass surface). Insert the comb about 1 cm from one end of the gel. If there are any bubbles in the agarose, remove them with a pipette.

5. Allow the gel to cool and solidify (about 20 minutes). Do not disturb it during this time. The agarose will become opaque when solidified.

6. When the agarose has solidified, place the gel in the electrophoresis chamber (if it is not already there) and add just enough 1× TAE Buffer to cover the gel to a depth of about 3–4 mm. Gently remove the comb, pulling it straight upwards, in one motion.

    Note: Agarose gels can be poured a few days in advance of their use and stored at 4°C. If the gel will be run within a day or two, it can be stored in the electrophoresis chamber covered with 1× TAE Buffer. If it will be longer than 3 days, store the gel wrapped in plastic wrap, since bacteria, capable of producing enzymes that degrade DNA, grow well in $1\times$ TAE Buffer. The gel (and buffer) should be brought to room temperature before being used.

Loading and Running Gels

1. Mix the samples of DNA with the appropriate volume of $2\times$ or $6\times$ DNA Gel Loading Buffer to give a final concentration of $1\times$ Loading Buffer (0.04% Bromophenol Blue, 8.3 mM EDTA, 6.7% sucrose). Pipette up and down to mix. Centrifuge briefly to collect the entire sample in the bottom of the tube.

2. Using a micropipette, withdraw the sample and load it into a well in the gel by lowering the pipette tip slightly into the well and slowly expelling the sample. Do not insert the tip too far into the well. If this is done, the well could be punctured, and the sample would be lost. Keep a record of which sample was loaded in each lane.

3. One well of each gel should be loaded with DNA molecular weight markers, which contain linear fragments of DNA of known sizes and masses.

4. Replace the cover of the electrophoresis chamber and connect the electrical leads to the power supply so that the negatively charged DNA will migrate to the positive electrode. The negative electrode should be at the well-end (top) of the gel, and the positive electrode should be at the bottom of the gel.

5. Turn on the power supply and set the voltage at ~90 volts. (If everything is connected properly, small bubbles should be visible at the electrodes in the chamber.) Run the gel until the blue dye has migrated the appropriate distance through the gel.

6. When finished, turn off the power supply and disconnect the electrical leads before opening the chamber and removing the gel.

Staining Gels with Ethidium Bromide

The DNA will be visualized by staining the gel with ethidium bromide. Ethidium bromide is a flat planar molecule that slides between the bases of DNA. ***ETHIDIUM BROMIDE IS A MUTAGEN AND A SUSPECTED CARCINOGEN. GLOVES SHOULD ALWAYS BE WORN WHEN HANDLING THIS CHEMICAL. THE WORK AREA SHOULD BE COVERED WITH PLASTIC-BACKED ABSORBENT PAPER.***

Use common sense when handling ethidium bromide. Once your gloved hand has touched an ethidium bromide solution or ethidium bromide-stained gel, do not use that hand to open doors, turn light switches, etc. It is best to handle the ethidium bromide-stained gel with one hand and use the other hand to open doors, turn light switches, etc. Better yet, work with your lab partner. One of you can wear gloves and handle the ethidium bromide stained-gel and ethidium bromide solutions, while the other remains "clean" to open doors, turn light switches, handle the Polaroid camera, take notes, etc.

Two methods of staining are presented in the following:

1. **Gels can be stained after electrophoresis.** The safest and easiest way to accomplish this is to stain the gel in a bag. Cut off the upper right-hand corner of the gel above the wells so that you will know when the gel is right side up. Use a plastic pancake turner to transfer the gel from the chamber to a quart-sized zippered bag. Add ~15 ml of 0.5 μg/ml ethidium bromide solution, and gently rock the bag with the gel for 20–25 minutes. You do not have to remove the gel (or the solution) from the bag

to view it. Just place the bag with the gel (right side up) on the UV trans-illuminator. Thus, your contact with ethidium bromide is minimized. ***WEAR EYE PROTECTION WHEN USING UV LIGHT.*** Use a Polaroid camera with an orange filter to photograph the gel. The camera setting will depend upon the type of film used. Discard the bag with the gel in a container for solid ethidium bromide waste.

2. **Gels can be stained during electrophoresis.** After the samples are loaded, add the dye to the running buffer to a final concentration of about 0.075 μg/ml. If your electrophoresis chamber holds ~250 ml of running buffer, add ~20 μl of a 1 mg/ml ethidium bromide solution. Place the ethidium bromide near the bottom of the chamber, and after the samples have migrated from the wells into the gel, the chamber can be rocked gently once or twice to distribute the dye more evenly.

   Upon completion of electrophoresis, discard the running buffer in a container for liquid ethidium bromide waste (which must be decontaminated before disposal). Cut off the upper right-hand corner of the gel, and transfer the gel from the chamber to a shallow container for transporting to the UV transilluminator. View and photograph the gel. Discard the gel in a container for solid ethidium bromide waste.

STERILE TECHNIQUE

- Wipe the lab bench with a disinfectant solution (or 10% chlorine bleach solution) before and after working with live bacteria.
- When you remove the lid or cap from a sterile bottle or plate, avoid placing it on the lab bench. Hold it facing downward using the thumb and little finger of your left hand. The rim and inside of the cap or lid should not touch any non-sterile surface.
- Pass the mouth of an open tube or bottle through a flame, holding it at an angle. Flaming warms the air at the opening and creates positive pressure, which prevents contaminants from falling into the container.
- When pipetting to or from tubes or bottles, hold them at an angle to prevent the introduction of contaminants.
- Place only sterile objects into a sterile tube or bottle. When using an adjustable-volume micropipette, remember that only the disposable tip is sterile; the rest of the pipette is not sterile.
- Work quickly. After completing the operation, flame the mouth of the tube or bottle again and replace the cap.
- When using a sterile serological pipette, touch only the large end opposite the tip. Hold the pipette firmly, about 1–2 inches from the large end, when inserting it into a pipette bulb or pump. If it is a glass pipette, draw the lower part of the pipette through a flame and insert only the untouched lower portion of the pipette into a sterile container.
- *Remember:* Any sterile object that comes into contact with a nonsterile surface or object is no longer sterile.

PROPER ENZYME USAGE

Enzymes, such as restriction endonucleases, DNA polymerases, and DNA ligases, are important tools in molecular biology and biotechnology. To ensure that they function optimally, it is important that they be handled properly. Commercial companies provide them at high concentrations in glycerol-based solutions to increase their stability during storage. Enzymes are generally stored at $-20°C$ and should be kept on ice when outside the freezer. Keeping them cold minimizes their inactivation. The following precautions should be followed when handling enzymes.

- When using enzymes, keep them on ice at all times. This means that you should not place a tube of enzyme in a rack at your workbench while you set up the rest of your reaction.
- Always centrifuge the tube briefly before opening the tube. Be careful not to touch the cap or the upper rim of the tube with your fingers. You do not want to introduce nucleases that could degrade the DNA in your reactions, and you do not want to introduce proteases that could degrade the enzyme.
- Hold the tube of enzyme with your fingers positioned below the rim but above the level of the enzyme solution. The heat from your fingers could rapidly warm up the enzyme solution. Holding the tube in this manner also allows you to visually monitor your pipetting. Hold the tube at eye level so you can watch as you put the pipette tip just below the surface of the solution and watch as the pipette tip fills with enzyme solution.
- The enzyme/glycerol solution is very viscous. You must pipette slowly and carefully to measure accurately.
- The activity of restriction enzymes can be affected by glycerol. The volume of enzyme/glycerol solution should never exceed 10% of the total reaction volume.
- Certain restriction enzymes can exhibit "star" activity under certain conditions. Star activity refers to an enzyme cutting sequences other than its normal recognition sequence. Star activity can be caused by high glycerol concentrations or low ionic strength of the buffer (i.e., the wrong buffer was used or the wrong amount was used).
- When assembling a reaction, always add the enzyme last (so it is placed in the correct concentration of buffer).
- Do not vortex enzyme mixtures. Hold the tube with one hand and flick it several times with your other hand. Centrifuge the tube briefly.

Bibliography

ORIGINAL RESEARCH ARTICLES

Altschul, S. F.; Boguski, M. S.; Gish, W.; and J. C. Wootton. 1994. Issues in searching molecular sequence databases. *Nature Genet.* 6:119–129.

Bernfeld, P. 1951. Enzymes of starch degradation and synthesis. In Advances *Enzymology, XII.* (F. Nord, ed.). New York: Interscience Publishing, 379.

Bernfeld, P. 1955. Amylases [α] and [β]. *Methods in Enzymology, I.* (S. Colowick and N. Kaplan, eds.). New York: Academic Press, 149.

Declerck, N.; Machius, M.; Wiegand, G.; Huber, R.; and C. Gaillardin. 2000. Probing structural determinants specifying high thermostability in *Bacillus licheniformis* α-amylase. *J. Mol. Biol.* 301:1041–1057.

Elhai, J., and C. P. Wolk. 1988. A versatile class of positive-selection vectors based on the nonviability of palindrome-containing plasmids that allows cloning into long polylinkers. *Gene* 68:119–138.

Hagan, C. E., and G. J. Warren. 1982. Lethality of palindromic DNA and its use in selection of recombinant plasmids. *Gene* 19:147–151.

MacGregor, E. A. 1988. α-amylase structure and activity. *J. Protein Chem.* 7:399–415.

Machius, M.; Weigand, G.; and R. Huber. 1995. Crystal structure of calcium-depleted *Bacillus licheniformis* α-amylase at 2.2 Å resolution. *J. Mol. Biol.* 246:545–559.

Mazur, A. K.; Haser, R.; and F. Payan. 1994. The catalytic mechanism of α-amylases based upon enzyme crystal structures and model building calculations. *Biochem. Biophys. Res. Commun.* 204:297–302.

Nielsen, J. E., and T. V. Borchert. 2000. Protein engineering of bacterial α-amylase. *Biochim. Biophys Acta.* 1543:253–274.

Qian, M.; Haser, R.; Buisson, G.; Duee, E.; and F. Payan. 1994. The active center of a mammalian α-amylase. Structure of the complex of a pancreatic α-amylase with a carbohydrate inhibitor refined to 2.2-Å resolution. *Biochemistry* 33:6284–6294.

Qian, M.; Spinelli, S.; Driguez, H.; and F. Payan. 1997. Structure of a pancreatic α-amylase bound to a substrate analogue at 2.03 Å resolution. *Protein Sci.* 6:2285–2296.

Schuler, G. D.; Altschul, S. F.; and D. J. Lipman. 1991. A workbench for multiple alignment construction and analysis. *Proteins* 9:180–190.

Sibakov, M., and I. Palva. 1984. Isolation and the 5′-end nucleotide sequence of *Bacillus licheniformis* α-amylase gene. *Eur. J. Biochem.* 145:567–572.

Stephens, M. A.; Ortlepp, S. A.; Ollington, J. F; and D. J. McConnell. 1984. Nucleotide sequence of the 5′ region of the *Bacillus licheniformis* α-amylase gene: Comparison with the *B. amyloliquefaciens* gene. *J. Bacteriol.* 158:369–372.

Tomazic, S. J., and A. M. Klibanov. 1988. Mechanisms of irreversible thermal inactivation of *Bacillus* α-amylases. *J. Biol. Chem.* 263:3086–3091.

Tomazic, S. J., and A. M. Klibanov. 1988. Why is one *Bacillus* α-amylase more resistant against irreversible thermoinactivation than another? *J. Biol. Chem.* 263:3092–3096.

Yuuki, T.; Nomura, T.; Tezuka, H.; Tsuboi, A.; Yamagata, H.; Tsukagoshi, N.; and S. Udaka. 1985. Complete nucleotide sequence of a gene coding for heat- and pH-stable α-amylase of *Bacillus licheniformis*: Comparison of the amino acid sequences of three bacterial liquefying α-amylases deduced from the DNA sequences. *J. Biochem.* 98:1147–1156.

REVIEW ARTICLES

*Cornelis P. 1987. Microbial amylases. *Microbiol. Sci.* 4:342–343.

Farber, G. K., and G. A. Petsko. 1990. The evolution of α/β barrel enzymes. *Trends Biochem. Sci.* 15:228–234.

Janecek, S. 1994. Parallel β/α-barrels of α-amylase, cyclodextrin glycosyltransferase and oligo-1,6-glucosidase versus the barrel of β-amylase: Evolutionary distance is a reflection of unrelated sequence. *FEBS Lett.* 353:119–123.

Janecek, S., and S. Balaz. 1992. α-amylases and approaches leading to their enhanced stability. *FEBS Lett.* 304:1–3.

*Pandey, A.; Nigam, P.; Soccol, C. R.; Soccol, V. T.; Singh, D.; and R. Mohan. 2000. Advances in microbial amylases. *Biotechnol. Appl. Biochem.* 31:135–152.

Svensson, B. 1994. Protein engineering in the α-amylase family: Catalytic mechanism, substrate specificity, and stability. *Plant. Mol. Biol.* 25:141–157.

LABORATORY MANUALS

Ausubel, F.; Brent, M. R.; Kingston, R. E.; Moore, D. D.; Seidman, J. G.; Smith, J. A.; and K. Struhl. 1987. *Current Protocols in Molecular Biology.* New York: Greene Publishing Associates and Wiley-Interscience.

Dieffenbach, C. W., and G. S. Dveksler. 1995. *PCR Primer. A Laboratory Primer.* Cold Spring Harbor, New York: Cold Spring Harbor Laboratory Press.

Harlow, E., and D. Lane. 1988. *Antibodies: A Laboratory Manual.* Cold Spring Harbor, New York: Cold Spring Harbor Laboratory Press.

Sambrook, J.; Fritsch, E. F.; and T. Maniatis. 1989. *Molecular Cloning: A Laboratory Manual.* Cold Spring Harbor, New York: Cold Spring Harbor Laboratory Press.

*Seidman, L. A., and C. Moore. 2000. *Basic Laboratory Methods for Biotechnology.* Upper Saddle River, N.J.: Prentice-Hall.

*Recommended reading for students.

BOOKS

Alberts, B.; Bray, D.; Lewis, J.; Raff, M.; Roberts, K.; and J. D. Watson. 1994. *Molecular Biology of the Cell.* New York: Garland Publishing.

Branden, C., and J. Tooze. 1991. *Introduction to Protein Structure.* New York: Garland Publishing.

*Chan, E. C. S.; Pelczar, M. G., Jr.; and N. R. Krieg. 1993. *Laboratory Exercises in Microbiology.* New York: McGraw-Hill.

*Clark, D. P., and L. D. Russell. 1997. *Molecular Biology Made Simple and Fun.* Vienna, Ill.: Cache River Press.

*Darby, N. J., and T. E. Creighton. 1993. *Protein Structure.* New York: IRL Press.

*Davies, J., and W. S. Reznikoff. 1992. *Milestones in Biotechnology.* Boston: Butterworth-Heinemann.

*Nicholl, S. T. 1994. *An Introduction to Genetic Engineering.* Cambridge, England: Cambridge Univ. Press.

Stryer, L. 1995. *Biochemistry.* New York: W. H. Freeman and Company.

*Watson, J. D.; Gilman, M.; Witkowski, J.; and M. Zoller. 1992. *Recombinant DNA.* New York: W. H. Freeman and Co.

*Weaver, R. F. 2002. *Molecular Biology.* New York: McGraw-Hill.

*Recommended reading for students.